Applied Mathematical Sciences

EDITORS

Fritz John
Courant Institute of Mathematical Sciences
New York University
New York, NY 10012

J.E. Marsden
Department of Mathematics
University of California
Berkeley, CA 94720

Lawrence Sirovich
Division of Applied Mathematics
Brown University
Providence, RI 02912

ADVISORS

H. Cabannes University of Paris-VI

M. Ghil New York University

J.K. Hale Brown University

J. Keller Stanford University

J.P. LaSalle Brown University

G.B. Whitham California Inst. of Technology

EDITORIAL STATEMENT

The mathematization of all sciences, the fading of traditional scientific boundaries, the impact of computer technology, the growing importance of mathematical-computer modelling and the necessity of scientific planning all create the need both in education and research for books that are introductory to and abreast of these developments.

The purpose of this series is to provide such books, suitable for the user of mathematics, the mathematician interested in applications, and the student scientist. In particular, this series will provide an outlet for material less formally presented and more anticipatory of needs than finished texts or monographs, yet of immediate interest because of the novelty of its treatment of an application or of mathematics being applied or lying close to applications.

The aim of the series is, through rapid publication in an attractive but inexpensive format, to make material of current interest widely accessible. This implies the absence of excessive generality and abstraction, and unrealistic idealization, but with quality of exposition as a goal.

Many of the books will originate out of and will stimulate the development of new undergraduate and graduate courses in the applications of mathematics. Some of the books will present introductions to new areas of research, new applications and act as signposts for new directions in the mathematical sciences. This series will often serve as an intermediate stage of the publication of material which, through exposure here, will be further developed and refined. These will appear in conventional format and in hard cover.

MANUSCRIPTS

The Editors welcome all inquiries regarding the submission of manuscripts for the series. Final preparation of all manuscripts will take place in the editorial offices of the series in the Division of Applied Mathematics, Brown University, Providence, Rhode Island.

SPRINGER-VERLAG NEW YORK INC., 175 Fifth Avenue, New York, N.Y. 10010

Printed in U.S.A.

Applied Mathematical Sciences | Volume 50

Applied Mathematical Sciences

1. John: **Partial Differential Equations**, 4th ed. (cloth)
2. Sirovich: **Techniques of Asymptotic Analysis.**
3. Hale: **Theory of Functional Differential Equations**, 2nd ed. (cloth)
4. Percus: **Combinatorial Methods.**
5. von Mises/Friedrichs: **Fluid Dynamics.**
6. Freiberger/Grenander: **A Short Course in Computational Probability and Statistics.**
7. Pipkin: **Lectures on Viscoelasticity Theory.**
8. Giacaglia: **Perturbation Methods in Non-Linear Systems.**
9. Friedrichs: **Spectral Theory of Operators in Hilbert Space.**
10. Stroud: **Numerical Quadrature and Solution of Ordinary Differential Equations.**
11. Wolovich: **Linear Multivariable Systems.**
12. Berkovitz: **Optimal Control Theory.**
13. Bluman/Cole: **Similarity Methods for Differential Equations.**
14. Yoshizawa: **Stability Theory and the Existence of Periodic Solutions and Almost Periodic Solutions.**
15. Braun: **Differential Equations and Their Applications**, 3rd ed. (cloth)
16. Lefschetz: **Applications of Algebraic Topology.**
17. Collatz/Wetterling: **Optimization Problems.**
18. Grenander: **Pattern Synthesis: Lectures in Pattern Theory, Vol I.**
19. Marsden/McCracken: **The Hopf Bifurcation and its Applications.**
20. Driver: **Ordinary and Delay Differential Equations.**
21. Courant/Friedrichs: **Supersonic Flow and Shock Waves.** (cloth)
22. Rouche/Habets/Laloy: **Stability Theory by Liapunov's Direct Method.**
23. Lamperti: **Stochastic Processes: A Survey of the Mathematical Theory.**
24. Grenander: **Pattern Analysis: Lectures in Pattern Theory, Vol. II.**
25. Davies: **Integral Transforms and Their Applications.**
26. Kushner/Clark: **Stochastic Approximation Methods for Constrained and Unconstrained Systems.**
27. de Boor: **A Practical Guide to Splines.**
28. Keilson: **Markov Chain Models—Rarity and Exponentiality.**
29. de Veubeke: **A Course in Elasticity.**
30. Sniatycki: **Geometric Quantization and Quantum Mechanics.**
31. Reid: **Sturmian Theory for Ordinary Differential Equations.**
32. Meis/Markowitz: **Numerical Solution of Partial Differential Equations.**
33. Grenander: **Regular Structures: Lectures in Pattern Theory, Vol. III.**
34. Kevorkian/Cole: **Perturbation Methods in Applied Mathematics.** (cloth)
35. Carr: **Applications of Centre Manifold Theory.**

(continued on inside back cover)

Calvin H. Wilcox

Sound Propagation in Stratified Fluids

Springer-Verlag
New York Berlin Heidelberg Tokyo

Calvin H. Wilcox
University of Utah
Department of Mathematics
Salt Lake City, Utah 84112
U.S.A.

AMS Classification: 46NO5, 76005

Library of Congress Cataloging in Publication Data
Wilcox, Calvin H. (Calvin Hayden)
Sound propagation in stratified fluids.
(Applied mathematical sciences ; v. 50)
Bibliography: p.
Includes index.
1. Sound-waves–Transmission. 2. Fluids–Acoustic properties. 3. Stratified flow. I. Title. II. Title: Stratified fluids. III. Series: Applied mathematical sciences (Springer-Verlag New York Inc.) ; v. 50.
QA1.A647 vol. 50 [QC233] 510s [534'.23] 84-1447

© 1984 by Springer-Verlag New York Inc.
All rights reserved. No part of this book may be translated or reproduced in any form without written permission from Springer-Verlag, 175 Fifth Avenue, New York, N.Y. 10010, U.S.A.

Printed and bound by R.R. Donnelley & Sons, Harrisonburg, Virginia.
Printed in the United States of America.

9 8 7 6 5 4 3 2 1

ISBN 0-387-90986-9 Springer-Verlag New York Berlin Heidelberg Tokyo
ISBN 3-540-90986-9 Springer-Verlag Berlin Heidelberg New York Tokyo

Preface

This monograph was begun during my sabbatical year in 1980, when I was a visiting professor at the University of Bonn, and completed at the University of Utah in 1981. Preliminary studies were carried out during the period 1972-79 at Utah and while I held visiting professorships at the University of Liège (1973-74), the University of Stuttgart (1974 and 1976-77) and the Ecole Polytechnique Fédérale of Lausanne (1979). Throughout this period my research was supported by the U.S. Office of Naval Research. I should like to express here my appreciation for the support of the Universities of Bonn, Liège, Stuttgart and Utah, the Ecole Polytechnique Fédérale, the Alexander von Humboldt Foundation and the Office of Naval Research which made the work possible. My special thanks are expressed to Professors S. Chatterji (Lausanne), H. G. Garnir (Liège), R. Leis (Bonn) and P. Werner (Stuttgart) for arranging my visits to their universities. I also want to thank Professor Jean Claude Guillot of the University of Paris who collaborated with me during the period 1975-78 on the spectral theory of the Epstein operator. That work, and our many discussions during that period of wave propagation in stratified media, contributed importantly to the final form of the work presented here.

Calvin H. Wilcox
Bonn
August, 1982

Introduction

Stratified fluids whose densities, sound speeds and other parameters are functions of a single depth coordinate occur widely in nature. Indeed, the earth's gravitational field imposes a stratification on its atmosphere, oceans and lakes. It is well known that their stratification has a profound effect on the propagation of sound in these fluids. The most striking effect is probably the occurrence of acoustic ducts, due to minima of the sound speed, that can trap sound waves and cause them to propagate horizontally. The reflection, transmission and distortion of sonar signals by acoustic ducts is important in interpreting sonar echoes. Signal scattering by layers of microscopic marine organisms is important to both sonar engineers and marine biologists. Again, reflection of signals from bottom sediment layers overlying a penetrable bottom are of interest both as sources of unwanted echoes and in the acoustic probing of such layers. Many other examples could be given.

The purpose of this monograph is to develop from first principles a theory of sound propagation in stratified fluids whose densities and sound speeds are essentially arbitrary functions of the depth. In physical terms, the propagation of both time-harmonic and transient fields is analyzed. The corresponding mathematical model leads to the study of boundary value problems for a scalar wave equation whose coefficients contain the prescribed density and sound speed functions. In the formalism adopted here these problems are intimately related to the spectral analysis of a partial differential operator, acting in a Hilbert space of functions defined in the domain occupied by the fluid.

The intended audience for this monograph includes both those applied physicists and engineers who are concerned with sound propagation in stratified fluids and those mathematicians who are interested in spectral analysis and boundary value problems for partial differential operators.

An attempt to address simultaneously two such disparate groups must raise the question: is there a common domain of discourse? The honest answer to this question is no! Current mathematical literature on spectral analysis and boundary value problems is based squarely on functional analysis, particularly the theory of linear transformations in Hilbert spaces. This theory has been readily accessible ever since the publication of M. H. Stone's AMS Colloquium volume in 1932. Nevertheless, the theory has not become a part of the curricula of applied physics and engineering and it is seldom seen in applied science literature on wave propagation. Instead, that literature is characterized by, on the one hand, the use of heuristic non-rigorous arguments and, on the other, by formal manipulations that typically involve divergent series and integrals, generalized functions of unspecified types and the like.

The differences in style and method outlined above pose a dilemma. Can an exposition of our subject be written that is accessible and useful to both applied scientists and mathematicians? An attempt is made to do this below by beginning each chapter with a substantial summary. Taken together, the summaries present the basic physical concepts and results of the theory, formulated in the simplest and most concise form consistent with their nature.

The purpose of the summaries is twofold. First, they can be interpreted in the heuristic way favored by applied physicists and engineers. When read in this way they are independent of the rest of the text and present a complete statement of the physical content of the theory. Second, readers conversant with Hilbert space theory can interpret the summaries as concise statements of the principal concepts and results of the rigorous mathematical theory. When read in this second way, the summaries serve as an introduction to and overview of the complete theory.

Contents

Chapter 1
Introduction

This monograph presents a theory of the propagation of transient sound waves in plane stratified fluids whose densities $\rho(y)$ and sound speeds $c(y)$ are functions of the depth y. The main goal of the theory is to calculate the signals produced by prescribed localized sources and to determine their asymptotic behavior for large times. The results include criteria for the occurrence of acoustic ducts that trap a portion of the signal and cause it to propagate horizontally.

The mathematical theory is based on a real valued function $u(t,X)$, the acoustic pressure or potential at the time t and the spatial point with Cartesian coordinates $X = (x_1, x_2, y)$, which satisfies a scalar wave equation that contains the fluid density $\rho(y)$ and sound speed $c(y)$. A derivation of this wave equation from the laws of fluid dynamics is presented here in order to clarify the hypotheses needed for its validity. Analogous derivations may be found in the monographs of F. G. Friedlander [6]* and L. M. Brekhovskihk [2] and in an article of I. Tolstoy [19].

A compressible, non-viscous, heat conducting fluid is considered. The states of the fluid are characterized by the variables

$$
(1.1) \qquad \begin{cases} p = \text{fluid pressure} \\ \rho = \text{fluid density} \\ \vec{v} = \text{fluid velocity} \\ T = \text{fluid temperature (absolute scale)} \\ S = \text{fluid entropy per unit mass} \end{cases}
$$

Each is a function of the variables t and X. These variables satisfy the

*Numbers in square brackets denote references from the list at the end of the monograph.

equations

$$\frac{D\rho}{Dt} + \rho \nabla \cdot \vec{v} = 0, \tag{1.2}$$

$$\rho \frac{D\vec{v}}{Dt} + \nabla p = \vec{G}, \tag{1.3}$$

$$\rho \frac{DS}{Dt} - \frac{1}{T} \nabla \cdot (k\nabla T) = 0, \tag{1.4}$$

$$f(p,\rho,T) = 0 \text{ and } \phi(p,\rho,S) = 0, \tag{1.5}$$

where

$$\frac{Du}{Dt} = \frac{\partial u}{\partial t} + \vec{v} \cdot \nabla u \tag{1.6}$$

denotes the material derivative. Equations (1.2) and (1.3) express the conservation of mass and linear momentum, respectively. The field $\vec{G} = \vec{G}(t,X)$ describes the external forces that act on the fluid. Equation (1.4) relates the variation of S to the diffusion of heat in a medium with thermal conductivity $k = k(X)$. Equations (1.5) are alternative forms of the thermodynamic equation of state.

Sound waves will be studied in a static inhomogeneous fluid in which external forces are negligible. Thus if the fluid parameters (1.1) for the static fluid are denoted by a subscript zero then one has

$$\vec{v}_0 = \vec{0}, \frac{\partial T_0}{\partial t} = 0, \frac{\partial S_0}{\partial t} = 0, \vec{G}_0 = \vec{0}, \tag{1.7}$$

whence

$$\frac{\partial \rho_0}{\partial t} = 0 \text{ or } \rho_0 = \rho_0(X) \tag{1.8}$$

$$\nabla p_0 = 0 \text{ or } p_0 = p_0(t) \tag{1.9}$$

and

$$\nabla \cdot (k\nabla T_0) = 0. \tag{1.10}$$

The last equation, together with suitable boundary conditions, determines a unique

$$T_0 = T_0(X). \tag{1.11}$$

Then (1.5) implies that

$$f(p_0(t),\rho_0(X),T_0(X)) = 0, \tag{1.12}$$

whence

$$p_0 = \text{const}. \tag{1.13}$$

and $\rho_0(X)$ may be determined by solving (1.12) with known p_0 and $T_0(X)$. Finally, $S_0(X)$ is determined by

$$\phi(p_0,\rho_0(X),S_0(X)) = 0. \tag{1.14}$$

The equations governing sound waves in a static inhomogeneous fluid will be obtained by a formal perturbation method applied to the static solution just discussed. The sound waves will be assumed to be excited by body forces with force density

$$\vec{G} = \varepsilon\, \vec{G}_1 \tag{1.15}$$

where ε is a parameter. Moreover, ε and $\vec{G}_1$ will be assumed to be so small that the resulting disturbance is adequately described by the first order terms in a Taylor series expansion in ε. The perturbation equations are obtained by substituting (1.15) and

$$\begin{cases} p = p_0 + \varepsilon p_1 \\ \rho = \rho_0 + \varepsilon\rho_1 \\ \vec{v} = \varepsilon\, \vec{v}_1 \\ T = T_0 + \varepsilon T_1 \\ S = S_0 + \varepsilon S_1 \end{cases} \tag{1.16}$$

in equations (1.2)-(1.5) and dropping terms containing powers of ε higher than the first.

Together with the small displacement assumption (1.16) the additional hypothesis will be made that the acoustic disturbances are so rapid that heat diffusion is negligible during passage of the sound waves. This assumption, which is called the adiabatic hypothesis, is equivalent to dropping the diffusion term in (1.4) by setting $k = 0$. Thus for sound waves one replaces (1.4) by

$$\frac{DS}{Dt} = 0. \tag{1.17}$$

Instead of linearizing (1.17) directly one may take the material derivative of the equation of state $\phi(p,\rho,S) = 0$ and use (1.17) to get

$$\phi_p \frac{Dp}{Dt} + \phi_\rho \frac{D\rho}{Dt} = 0 \tag{1.18}$$

where $\phi_p = \partial\phi/\partial p$ and $\phi_\rho = \partial\phi/\partial\rho$. Substituting (1.16) in (1.18) and retaining only linear terms in ε gives (since $\nabla p_0 = 0$)

$$\phi_p^0 \frac{\partial p_1}{\partial t} + \phi_\rho^0 \left(\frac{\partial \rho_1}{\partial t} + \nabla \rho_0 \cdot \vec{v}_1\right) = 0 \tag{1.19}$$

where

$$\begin{cases} \phi_p^0 = \phi_p(p_0, \rho_0(X), S_0(X)), \\ \phi_\rho^0 = \phi_\rho(p_0, \rho_0(X), S_0(X)). \end{cases} \tag{1.20}$$

This can be written

$$\frac{\partial p_1}{\partial t} = c^2(X)\left(\frac{\partial \rho_1}{\partial t} + \nabla \rho_0 \cdot \vec{v}_1\right) \tag{1.21}$$

where

$$c^2(X) = -\frac{\phi_\rho^0}{\phi_p^0} = \left(\frac{\partial p}{\partial \rho}\right)_S \tag{1.22}$$

is the local speed of sound at X. Linearizing equations (1.2) and (1.3) yields

$$\frac{\partial \rho_1}{\partial t} + \nabla \cdot (\rho_0 \vec{v}_1) = 0, \tag{1.23}$$

$$\rho_0 \frac{\partial \vec{v}_1}{\partial t} + \nabla p_1 = \rho_0 \vec{G}_1. \tag{1.24}$$

Combining (1.21) and (1.23) gives the alternative equation

$$\frac{\partial p_1}{\partial t} + c^2 \rho_0 \nabla \cdot \vec{v}_1 = 0. \tag{1.25}$$

Finally, differentiating (1.25) with respect to t and using (1.24) to eliminate $\vec{v}_1$ gives a scalar wave equation for p_1:

$$\frac{\partial^2 p_1}{\partial t^2} - c^2\rho_0 \nabla \cdot \left(\frac{1}{\rho_0} \nabla p_1\right) = F \tag{1.26}$$

where

$$F = -c^2\rho_0 \nabla \cdot \vec{G}_1 \tag{1.27}$$

is determined by the prescribed source density $\vec{G}_1$.

If $\rho_0\vec{G}_1 = \nabla G$ then one may introduce an acoustic potential $u(t,X)$ such that

$$\vec{v}_1 = \frac{1}{\rho_0} \nabla u, \quad p_1 = -\frac{\partial u}{\partial t} + G. \tag{1.28}$$

Then (1.24) and (1.25) are satisfied if

$$\frac{\partial^2 u}{\partial t^2} - c^2\rho_0 \nabla \cdot \left(\frac{1}{\rho_0} \nabla u\right) = F \tag{1.29}$$

with $F = \partial G/\partial t$. It will be convenient to use this formulation below.

A static fluid will be said to be plane stratified if the fluid parameters depend on the single depth coordinate y. In this case, writing $c = c(y)$, $\rho_0 = \rho(y)$ in (1.29) one gets the partial differential equation

$$\frac{\partial^2 u}{\partial t^2} - c^2(y)\left[\frac{\partial^2 u}{\partial x_1^2} + \frac{\partial^2 u}{\partial x_2^2} + \rho(y) \frac{\partial}{\partial y}\left(\frac{1}{\rho(y)} \frac{\partial u}{\partial y}\right)\right] = F(t,x_1,x_2,y). \tag{1.30}$$

Sound waves governed by (1.30) are studied in the remainder of this monograph. The fluid media are assumed to be unlimited in horizontal planes, so that $-\infty < x_1,x_2 < \infty$. The principal case treated is that of a completely unlimited fluid for which $-\infty < y < \infty$. This idealized problem provides a model for the analysis of sound propagation in portions of a finite stratified fluid that are far from the boundaries. The cases of a semi-infinite layer, with $0 < y < \infty$, and a finite layer, with $0 < y < h < \infty$, are also discussed. These cases provide models for the study of sound propagation near one and two horizontal plane boundaries, respectively.

The fluid parameters $\rho(y)$ and $c(y)$ are inherently positive. It will be assumed here that they are bounded and bounded away from zero. This suffices to show the solvability of the propagation problem. Additional conditions that restrict the behavior of $\rho(y)$ and $c(y)$ at $y = \pm\infty$ are needed to obtain more detailed information concerning the structure of the acoustic field. Such conditions are imposed as they are needed.

The method of solving the wave equation (1.30) developed below is based on the spectral theory of selfadjoint operators in a Hilbert space. The possibility of using this method is suggested by the observation that the partial differential operator A defined by

$$Au = -c^2(y)\left[\frac{\partial^2 u}{\partial x_1^2} + \frac{\partial^2 u}{\partial x_2^2} + \rho(y)\frac{\partial}{\partial y}\left(\frac{1}{\rho(y)}\frac{\partial u}{\partial y}\right)\right] \tag{1.31}$$

is formally selfadjoint with respect to the scalar product defined by the integral

$$(u,v) = \iiint_\Omega \overline{u(x_1,x_2,y)}\, v(x_1,x_2,y)\, c^{-2}(y)\, \rho^{-1}(y)dx_1dx_2dy, \tag{1.32}$$

where $c^{-2}(y) = 1/c^2(y)$, $\rho^{-1}(y) = 1/\rho(y)$ and the integration is over the domain Ω occupied by the fluid. Indeed, partial integration using the divergence theorem yields

$$(Au,v) = (u,Av) + \text{boundary terms} \tag{1.33}$$

and if u and v satisfy suitable boundary conditions then one has $(Au,v) = (u,Av)$. These facts are used below to construct a selfadjoint realization of A in a Hilbert space $\mathcal{H}$ of functions with scalar product (1.32). This makes it possible to solve the wave equation (1.30) by means of the functional calculus based on the spectral theorem for selfadjoint operators. A concise discussion of the principal concepts and results of this theory will be presented here in order to motivate and clarify the analysis that follows. Detailed developments of the spectral theory of operators in Hilbert spaces can be found in a number of excellent texts, including those of N. Dunford and J. T. Schwartz [5] and T. Kato [11].

The set of all complex-valued functions u on a domain Ω in three-dimensional space that satisfy

$$\|u\|^2 = (u,u) = \iiint_\Omega |u(x_1,x_2,y)|^2\, c^{-2}(y)\, \rho^{-1}(y)dx_1dx_2dy < \infty \tag{1.34}$$

form a Hilbert space $\mathcal{H}$; i.e., a vector space over the complex field C that has a scalar product (1.32) and is complete in the norm $\|u\|$ defined by (1.34). The completeness property plays the same fundamental role in functional analysis that the completeness of the real number field plays in calculus. To obtain it one must include in $\mathcal{H}$ the large class of all Lebesgue square-integrable functions. The completeness of such spaces was discovered and proved by F. Riesz and E. Fischer, working independently, in 1907.

The differential operator A defined by (1.31) can be used to define various linear operators in $\mathcal{H}$. Of course, the domain of definition $D(A)$ of such an operator cannot be all of $\mathcal{H}$. Instead, functions $u \in D(A)$ must have the necessary derivatives and perhaps satisfy integrability and boundary conditions. Operators with different domains are considered to be different operators, even if the actions of both are defined by the same operator (1.31).

The operators considered here have domains $D(A)$ that, although they are rather "thin" subsets of $\mathcal{H}$, nevertheless are large enough to be dense in $\mathcal{H}$; i.e., every $u \in \mathcal{H}$ is a limit in the norm (1.34) of a sequence $u_n \in D(A)$. Such operators are said to be densely defined. Every such operator A has a well-defined adjoint operator A^* whose domain $D(A^*)$ is the set of all $u \in \mathcal{H}$ such that

$$(1.35) \qquad (Av,u) = (v,u^*) \text{ for all } v \in D(A).$$

The action of A^* is defined by $A^*u = u^*$.

The selfadjoint operators A are those for which $A^* = A$; i.e., $D(A^*) = D(A)$ and $A^*u = Au$ for all $u \in D(A)$. The importance of this rather special class of operators is due to the discovery of J. von Neumann in 1929 that they are precisely the operators that have a spectral representation

$$(1.36) \qquad A = \int_{-\infty}^{\infty} \lambda \, d\Pi(\lambda)$$

where $\{\Pi(\lambda) \mid -\infty < \lambda < \infty\}$ is a family of operators in $\mathcal{H}$, called the spectral family of A.

To explain the properties of spectral families and the meaning of (1.36), the notions of subspace and orthogonal projection in a Hilbert space must be defined. A subspace $M \subset \mathcal{H}$ is a subset that is closed under linear operations in $\mathcal{H}$ and is complete. Thus a subspace is itself a Hilbert space. To each subspace M there corresponds a second subspace $M^{\perp}$, called the orthogonal complement of M, which consists of all $u \in \mathcal{H}$ such that $(u,v) = 0$ for all $v \in M$. Moreover, every $u \in \mathcal{H}$ has a unique decomposition

$$u = u_1 + u_2,\ u_1 \in M,\ u_2 \in M^{\perp}. \tag{1.37}$$

Given a subspace M, the correspondence that assigns to each $u \in \mathcal{H}$ its component $u_1 \in M$ defines a linear operator Π in $\mathcal{H}$. It has the properties

$$\Pi u = u_1 \in M \text{ for all } u \in \mathcal{H}, \tag{1.38}$$

$$\Pi^2 = \Pi \tag{1.39}$$

$$\Pi^* = \Pi. \tag{1.40}$$

Conversely, any operator Π with properties (1.39), (1.40) defines a subspace $M = \Pi\mathcal{H}$. Such operators are called orthogonal projections. Thus there is a one-to-one correspondence between the set of subspaces of $\mathcal{H}$ and the set of all orthogonal projections in $\mathcal{H}$.

A spectral family $\{\Pi(\lambda) \mid -\infty < \lambda < \infty\}$ in $\mathcal{H}$ is a family of orthogonal projections $\Pi(\lambda)$ in $\mathcal{H}$ with the properties

$$\Pi(\lambda)\ \Pi(\mu) = \Pi(\mu) \text{ for all } \lambda \geq \mu \tag{1.41}$$

$$\left\{ \begin{array}{l} \lim\limits_{\lambda \to -\infty} \Pi(\lambda)u = 0 \\ \lim\limits_{\lambda \to +\infty} \Pi(\lambda)u = u \end{array} \right\} \text{ for } u \in \mathcal{H}. \tag{1.42}$$

It follows that $\{\Pi(\lambda)\}$ is a monotone non-decreasing family in the sense that

$$(\Pi(\mu)u,u) \leq (\Pi(\lambda)u,u) \text{ for all } \lambda \geq \mu \tag{1.43}$$

and all $u \in \mathcal{H}$. It is also customary to normalize $\{\Pi(\lambda)\}$ by requiring continuity from the right:

$$\Pi(\lambda+0)u = \Pi(\lambda)u \text{ for all } \lambda \in R \text{ and } u \in \mathcal{H}, \tag{1.44}$$

but this is not essential.

The spectral integral

$$\int_{-\infty}^{\infty} f(\lambda)\ d\Pi(\lambda), \tag{1.45}$$

where $f(\lambda)$ is a continuous complex-valued function of $\lambda \in R$, will be interpreted as a linear operator in $\mathcal{H}$. It will be defined in two steps. First, if $-\infty < a < b < \infty$ then

$$(1.46) \qquad \int_a^b f(\lambda)\, d\Pi(\lambda)u = \lim_{|\pi|\to 0} \sum_{j=1}^{n} f(\lambda_j)[\Pi(\lambda_j) - \Pi(\lambda_{j-1})]u$$

for all $u \in \mathcal{H}$, where $\pi : a = \lambda_0 < \lambda_1 < \cdots < \lambda_n = b$ is a partition of $[a,b]$ and $|\pi| = \operatorname{Max}_j (\lambda_j - \lambda_{j-1})$. The existence in $\mathcal{H}$ of the limit in (1.46) follows from the monotonicity property (1.43), just as in the case of scalar Stieltjes integrals. Finally

$$(1.47) \qquad \int_{-\infty}^{\infty} f(\lambda)\, d\Pi(\lambda)u = \lim_{M,N\to\infty} \int_{-N}^{M} f(\lambda)\, d\Pi(\lambda)u,$$

where convergence is in $\mathcal{H}$. The domain of the operator so defined is the set of $u \in \mathcal{H}$ for which the limit in (1.47) exists. It can be shown to coincide with the set of $u \in \mathcal{H}$ such that

$$(1.48) \qquad \int_{-\infty}^{\infty} |f(\lambda)|^2\, d\|\Pi(\lambda)u\|^2 < \infty.$$

In particular, the domain of the operator A characterized by (1.36) is the set of all $u \in \mathcal{H}$ such that

$$(1.49) \qquad \int_{-\infty}^{\infty} \lambda^2\, d\|\Pi(\lambda)u\|^2 < \infty.$$

As examples the spectral families associated with two special cases of the operator (1.31) will be described. In both cases $\rho(y) \equiv c(y) \equiv 1$ and $A = -\Delta$ is the negative Laplacian. In the first example Ω is a bounded domain in three-dimensional space and $D(A)$ is the subset of all $u \in \mathcal{H}$ for which $Au \in \mathcal{H}$ and $u(x) = 0$ on the boundary of Ω. Then A is selfadjoint and has a pure point spectrum consisting of real eigenvalues $0 < \lambda_1 \le \lambda_2 \le \cdots$ and corresponding eigenfunctions $u_1(x), u_2(x), \cdots$ such that $Au_n = \lambda_n u_n$ (i.e. $u_n \in D(A)$ and $Au_n = -\Delta u_n = \lambda_n u_n$). If the eigenfunctions are normalized; i.e., $\|u_n\| = 1$ for $n = 1,2,\cdots$, then the spectral family of A is given by

$$(1.50) \qquad \Pi(\lambda)u = \sum_{\lambda_n \le \lambda} \lambda_n (u_n,u)u_n$$

where the notation indicates a sum over all indices n such that $\lambda_n \le \lambda$. Thus in this case $\{\Pi(\lambda)\}$ is a step function that is constant for λ between consecutive eigenvalues. Note that the jump $\Pi(\lambda_n) - \Pi(\lambda_n - 0)$ is precisely the orthogonal projection onto the eigenspace of A for the eigenvalue λ_n.

For the second example consider $A = -\Delta$, acting on functions defined in the whole space $\Omega = R^3$. If D(A) is the subset of all $u \in \mathcal{H}$ for which $Au \in \mathcal{H}$ then A is selfadjoint and has a pure continuous spectrum consisting of all real $\lambda \geq 0$. The spectral family $\{\Pi(\lambda)\}$ may be constructed by Fourier analysis. Indeed, if $u \in \mathcal{H}$ then the limit

$$\hat{u}(P) = \lim_{R\to\infty} \frac{1}{(2\pi)^{3/2}} \int_{|X|\leq R} e^{-iP\cdot X} u(X)dX \tag{1.51}$$

exists in $\mathcal{H}$ and one has

$$u(X) = \lim_{R\to\infty} \frac{1}{(2\pi)^{3/2}} \int_{|P|\leq R} e^{iP\cdot X} \hat{u}(P)dP, \tag{1.52}$$

also in $\mathcal{H}$, and

$$\|\hat{u}\| = \|u\|. \tag{1.53}$$

These results are Plancherel's theory of the Fourier transform. With this notation the spectral family of $A = -\Delta$ in R^3 is given by

$$\Pi(\lambda)u(x) = \begin{cases} \dfrac{1}{(2\pi)^{3/2}} \displaystyle\int_{|P|^2\leq\lambda} e^{iP\cdot X} \hat{u}(P)dP, & \lambda \geq 0, \\ 0 & , \lambda \leq 0. \end{cases} \tag{1.54}$$

It is clear that in this case $\Pi(\lambda)$ varies continuously with λ.

The differential operator (1.31), acting in the whole space R^3 or in a semi-infinite or finite layer, is much closer to the second example than the first. In particular, it always has a pure continuous spectrum. However, when $\rho(y)$ and $c(y)$ are variable the structure of $\{\Pi(\lambda)\}$ may be complicated by the existence of acoustic ducts and corresponding families of subspaces of $\mathcal{H}$. Of course, the spectral resolution (1.34) is useful only when an explicit construction of $\{\Pi(\lambda)\}$, such as (1.50) or (1.54), can be found. This is achieved below for sound waves in stratified fluids by means of the physically motivated concept of normal modes.

The organization of the remainder of the monograph will now be described. Chapter 2 presents a construction of a selfadjoint realization of the acoustic propagator A defined by (1.31), followed by a solution of the wave equation (1.30) by means of the functional calculus (1.47), (1.48) for A.

The principal results of the monograph are developed in the long Chapter 3 which presents an explicit construction of the spectral family of A, under the condition that $\rho(y)$ and $c(y)$ have limits in a suitable sense at $y = \pm\infty$. The construction is based on two families of generalized eigenfunctions, or normal mode functions. Physically, the first family describes the response of the fluid to an incident plane acoustic wave. The second family exists only when the fluid has an acoustic duct and represents guided waves that are localized near the duct.

Chapter 4 presents a detailed analysis of the large-time behavior of transient sound fields in stratified fluids. The calculations begin with a normal mode expansion of the fields. Then, under suitable restrictions on the behavior of $\rho(y)$ and $c(y)$ at $y = \pm\infty$, the large-time behavior of the fields is determined by the method of stationary phase. The convergence of u and its first-order derivatives in $\mathcal{H}$ to the corresponding asymptotic wave functions is demonstrated and used to determine the asymptotic distribution of energy for large times.

Chapter 5 presents an analysis of the scattering of acoustic signals by an inhomogeneous stratified layer of finite thickness that separates two homogeneous fluids. A scattering operator is constructed and shown to determine completely the asymptotic form of the reflected and transmitted signals that are produced by a prescribed source localized in one of the homogeneous fluids.

The monograph concludes with two appendices. Appendix 1 presents in a concise form, with references to the literature, the Weyl-Kodaira-Titchmarsh theory of singular Sturm-Liouville operators. These results provide the foundation for the proof in Chapter 3 of the completeness and orthogonality of the normal mode functions. Appendix 2 presents a version of the stationary phase estimate for oscillatory integrals containing parameters. The uniformity with respect to the parameters of these estimates is essential for the applications in Chapter 4.

The physical literature on sound propagation in stratified fluids is very large. The modern developments date from the pioneering work of C. L. Pekeris [17], published in 1948, in which $\rho(y)$ and $c(y)$ are piecewise constant. In the intervening years hundreds of papers on the subject have appeared in the Journal of the Acoustical Society of America alone. Some of the material has been reviewed in the well known books of L. M. Brekhovskikh [2] and I. Tolstoy and C. S. Clay [20].

The work presented here differs from the earlier literature in being based on concepts and results from the theory of linear transformations in Hilbert spaces. Some of the many advantages of this approach will be mentioned here. First, the use of the Weyl-Kodaira-Titchmarsh theory of singular Sturm-Liouville operators permits a straightforward derivation of the completeness and orthogonality of the normal mode functions, without *ad hoc* hypotheses concerning $\rho(y)$ and $c(y)$. This theory is intrinsically a Hilbert space theory. Second, the use of the notions of convergence in the mean-square and energy norms leads to results on the asymptotic behavior for $t \to \infty$ of transient waves that are not true in the conventional sense of pointwise convergence. Finally, the introduction of the unitary scattering operator in Chapter 4 gives a particularly lucid form to the relationships among the incident, reflected and transmitted waveforms in the scattering of signals by an inhomogeneous layer.

Earlier applications of Hilbert space methods to wave propagation in stratified media were made by the author to the special case of the Pekeris profile [23,26,27,28] and by J. C. Guillot and Wilcox [7,8] and Y. Dermenjian, Guillot and Wilcox [4] to the special case of the Epstein profile.

For simplicity, displayed equations are numbered separately in each chapter. The notation (j.k) always means equation k of section j of the current chapter. The notation (i.j.k) is used to refer to equation k of section j of chapter i.

Chapter 2
The Propagation Problems and Their Solutions

The purpose of this chapter is to present a mathematical formulation and solution of the propagation problems that are studied in the subsequent chapters. The basic physical concepts and results are summarized in §1. The remaining sections develop the mathematical theory, together with rigorous proofs or references to the literature. The case of an unlimited fluid is treated here. The changes needed to treat semi-infinite and finite layers are indicated at the end of Chapter 3.

§1. SUMMARY

The acoustic field in an unlimited stratified fluid will be characterized by a real-valued acoustic potential $u(t,X)$ where $X = (x,y) = (x_1,x_2,y) \in R^3$ denote the rectangular spatial coordinates of Chapter 1. The partial derivatives of u will be denoted by $D_k u$, $k = 0, 1, 2, 3$, where

$$D_0 = \partial/\partial t, \; D_1 = \partial/\partial x_1, \; D_2 = \partial/\partial x_2, \; D_3 = \partial/\partial y. \tag{1.1}$$

With this notation the wave equation of Chapter 1 takes the form

$$D_0^2 u - c^2(y)\, \rho(y)\, \nabla \cdot \left(\frac{1}{\rho(y)} \nabla u\right) = F(t,X) \tag{1.2}$$

where $\nabla = (D_1,D_2,D_3)$ denotes the spatial gradient operator and $F(t,X)$ is a source density function. The acoustic pressure p and velocity field $\vec{v}$ in the fluid are given by [6]

$$p = -D_0 u \text{ and } \vec{v} = \frac{1}{\rho(y)} \nabla u \tag{1.3}$$

respectively.

As yet, no conditions other than positivity have been placed on the variations of $\rho(y)$ and $c(y)$. In the remainder of this work it is assumed as a minimal hypothesis that

$$(1.4) \qquad 0 < \rho_m \le \rho(y) \le \rho_M < \infty \text{ and } 0 < c_m \le c(y) \le c_M < \infty$$

for all y, where ρ_m, ρ_M, c_m and c_M are suitable constants. In this chapter no additional hypotheses are needed (other than the technical hypothesis of measurability). In particular, $\rho(y)$ and $c(y)$ need not be continuous. The meaning of the wave equation (1.2) when $\rho(y)$ and $c(y)$ are discontinuous is explained below. Other conditions than (1.4) that would permit $\rho(y)$ and $c(y)$ to be unbounded or to tend to zero can be treated by the methods developed here, but (1.4) is sufficiently general for most applications.

The wave sources are assumed to be localized in both space and time. Without loss of generality it can be assumed that the source density in (1.2) satisfies

$$(1.5) \qquad \text{supp } F \subset \{(t,X) : T \le t \le 0,\ x_1^2 + x_2^2 + (y - y_0)^2 \le \delta_0^2\}$$

where supp F denotes the support of F (i.e., the smallest closed set outside which F is zero). Then $u(t,X)$ can be characterized as the solution of (1.2) for $t \in R$, $X \in R^3$ that satisfies the condition $u(t,X) \equiv 0$ for all $t < T$. It is known that (1.4) implies that the supports of solutions of (1.2) propagate with speed not exceeding c_M [22]. Hence, (1.5) implies that if

$$(1.6) \qquad u(0,X) = f(X),\ D_0 u(0,X) = g(X) \text{ for } X \in R^3$$

then

$$(1.7) \qquad \text{supp } f \cup \text{supp } g \subset \{X : x_1^2 + x_2^2 + (y - y_0)^2 \le \delta^2\}$$

where $\delta = \delta_0 + c_M|T|$. It follows that $u(t,X)$ can be characterized for $t \ge 0$ as the solution of the homogeneous wave equation

$$(1.8) \qquad D_0^2 u - c^2(y)\, \rho(y)\, \nabla \cdot \left(\frac{1}{\rho(y)} \nabla u\right) = 0 \text{ for } t > 0,\ X \in R^3$$

with initial values f, g that satisfy (1.6), (1.7). The reduction of the initial value problem for (1.2) to that for (1.8) by means of Duhamel's principle is shown below.

The integral

$$(1.9) \qquad E(u,K,t) = \int_K \{|\nabla u(t,X)|^2 + c^{-2}(y)\ |D_0 u(t,X)|^2\}\ \rho^{-1}(y)\ dX,$$

where $dX = dx_1 dx_2 dy$, may be interpreted as the energy of the acoustic field u in the set $K \subset R^3$ at time t. It is known that solutions of (1.8) satisfy the conservation law

$$(1.10) \qquad E(u,R^3,t) = E(u,R^3,0)$$

where $E(u,R^3,0) \leq +\infty$. Only solutions with finite total energy are considered below. A necessary and sufficient condition for u to have this property is that the initial state have finite energy [22]

$$(1.11) \qquad \int_{R^3} \{|\nabla f(X)|^2 + c^{-2}(y)\ |g(X)|^2\}\ \rho^{-1}(y)\ dX < \infty.$$

The initial value problem in its classical formulation (1.6), (1.8) will have a solution only if $\rho(y)$, $c(y)$, $f(X)$ and $g(X)$ are sufficiently smooth. However, for arbitrary $\rho(y)$, $c(y)$ satisfying (1.4) the problem may be shown to have a unique solution with finite energy whenever the initial state f,g has this property. A proof is outlined in §3 below. A formal construction of the solution may be based on the operational calculus for the linear operator A, defined by the differential operator

$$(1.12) \qquad Au = -c^2(y)\ \rho(y)\ \nabla \cdot \left(\frac{1}{\rho(y)}\ \nabla u\right)$$

acting in the Hilbert space $\mathcal{H}$ with scalar product

$$(1.13) \qquad (u,v) = \int_{R^3} \overline{u(X)}\ v(X)\ c^{-2}(y)\ \rho^{-1}(y)\ dX.$$

If the domain of A is defined to be the set of all $u \in \mathcal{H}$ such that $\nabla u \in \mathcal{H}$ and $\nabla \cdot (\frac{1}{\rho} \nabla u) \in \mathcal{H}$ then A is a selfadjoint non-negative operator. Moreover,

$$(1.14) \qquad u(t,\cdot) = (\cos t\ A^{1/2})f + (A^{-1/2} \sin t\ A^{1/2})g$$

is the solution of the initial value problem

$$(1.15) \qquad \begin{cases} D_0^2 u + Au = 0,\ t > 0, \\ u(0) = f,\ D_0 u(0) = g. \end{cases}$$

It can be shown that (1.14) is the solution with finite energy whenever the initial values have finite energy.

The initial value problem for (1.2) where F satisfies (1.5) and $u(t,X) \equiv 0$ for $t < T$ can be reformulated as

$$(1.16) \qquad \begin{cases} D_0^2 u + Au = F(t,\cdot), \ t \in R, \\ u(t) = 0 \text{ for } t < T. \end{cases}$$

The solution is given by the Duhamel integral

$$(1.17) \qquad u(t,\cdot) = \int_T^t \{A^{-1/2} \sin (t-\tau)A^{1/2}\} \ F(\tau,\cdot) \ d\tau, \ t \geq T.$$

It is clear that for $t \geq 0$ this solution satisfies (1.15) with

$$(1.18) \qquad \begin{cases} f = u(0) = -\int_T^0 \{A^{-1/2} \sin \tau \ A^{1/2}\} \ F(\tau,\cdot) \ d\tau, \\ g = D_0 u(0) = \int_T^0 \{\cos \tau \ A^{1/2}\} \ F(\tau,\cdot) \ d\tau. \end{cases}$$

The remainder of the analysis will be based on the representation (1.14). It will be convenient to rewrite it in the form

$$(1.19) \qquad u(t,X) = \text{Re } \{v(t,X)\}$$

where $v(t,X)$ is the complex-valued wave function defined by

$$(1.20) \qquad v(t,\cdot) = e^{-itA^{1/2}} h$$

and

$$(1.21) \qquad h = f + i \ A^{-1/2} g \ .$$

This representation is valid if f and g are real-valued and $A^{1/2} f$, f, g and $A^{-1/2} g$ are in $\mathcal{H}$. A rigorous interpretation of (1.14)-(1.21) may be based on the calculus of selfadjoint operators in Hilbert spaces.

It is noteworthy that the same formalism is applicable to the cases of semi-infinite and finite layers. For the semi-infinite layer $R_+^3 = \{X : y > 0\}$, $\mathcal{H}$ is the Hilbert space with scalar product

$$(1.22) \qquad (u,v) = \int_{R^3_+} \overline{u(X)}\, v(X)\, c^{-2}(y)\, \rho^{-1}(y)\, dX$$

and the domain of A is the set of all $u \in \mathcal{H}$ such that $\nabla u \in \mathcal{H}$, $\nabla\cdot(\frac{1}{\rho}\nabla u) \in \mathcal{H}$ and u satisfies the Dirichlet or Neumann condition at $y = 0$. The relations (1.14)-(1.21) then hold as before. The analogous statements hold for the finite layer $R^3_h = \{X : 0 < y < h\}$.

§2. THE ACOUSTIC PROPAGATOR

In this section a precise definition of the acoustic propagator A is given and a proof of its selfadjointness in $\mathcal{H}$ is outlined. Here and throughout this work $\rho(y)$ and $c(y)$ are assumed to be Lebesgue measurable functions that satisfy the boundedness conditions (1.4). It follows that

$$(2.1) \qquad m(K) = \int_K c^{-2}(y)\, \rho^{-1}(y)\, dX$$

defines a measure in R^3 that is equivalent to Lebesgue measure (i.e. m is absolutely continuous with respect to dX and $dm/dX = c^{-2}(y)\, \rho^{-1}(y)$). Hence the Lebesgue space

$$(2.2) \qquad \mathcal{H} = L_2(R^3, c^{-2}(y)\, \rho^{-1}(y)\, dX)$$

is a Hilbert space with scalar product (1.13). Note that (1.4) implies that $\mathcal{H}$ is equivalent as a normed space to the usual Lebesgue space $L_2(R^3, dX)$, although they are distinct as Hilbert spaces.

A selfadjoint realization of the differential operator A in $\mathcal{H}$, also to be denoted by A, is obtained by defining the domain of A to be

$$(2.3) \qquad D(A) = L_2^1(R^3) \cap \{u : \nabla\cdot(\rho^{-1}\nabla u) \in L_2(R^3)\}$$

where $L_2(R^3) = L_2(R^3, dX)$ and

$$(2.4) \qquad L_2^1(R^3) = L_2(R^3) \cap \{u : \nabla u \in L_2(R^3)\}$$

is the usual first Sobolev space [1]. All the differential operations in (2.3), (2.4) are to be understood in the sense of the theory of distributions. It follows that the linear operator A in $\mathcal{H}$ defined by (1.12), (2.3) satisfies

$$(2.5) \qquad A = A^* \geq 0$$

where A^* is the adjoint of A with respect to the scalar product (1.13). A proof of (2.5) may be given by the method employed in [26]. Alternatively, (2.5) may be derived from Kato's theory of sesquilinear forms in Hilbert space [11, p. 322]. Indeed, if one defines a sesquilinear form $\mathcal{A}$ in $\mathcal{H}$ by

$$D(\mathcal{A}) = L_2^1(R^3) \subset \mathcal{H} \tag{2.6}$$

and

$$\mathcal{A}(u,v) = \int_{R^3} \nabla\bar{u} \cdot \nabla v \, \rho^{-1}(y) \, dX \tag{2.7}$$

then it is easy to verify that $\mathcal{A}$ is closed and non-negative, and that A is the unique selfadjoint non-negative operator in $\mathcal{H}$ associated with $\mathcal{A}$. As an additional dividend it follows from Kato's second representation theorem [11, p. 331] that $D(A^{1/2}) = L_2^1(R^3)$ and for all $u \in D(A^{1/2})$ one has

$$\|A^{1/2} u\|_{\mathcal{H}}^2 = \mathcal{A}(u,u) = \int_{R^3} |\nabla u|^2 \rho^{-1}(y) \, dX \tag{2.8}$$

where $\|\cdot\|_{\mathcal{H}}$ is the norm in $\mathcal{H}$.

§3. SOLUTIONS WITH FINITE ENERGY

It was shown in [22] that the initial value problem (1.6), (1.8) has a unique generalized solution with finite energy (= solution wFE) whenever the initial state f, g has finite energy; i.e., (1.11) holds. These conditions may be written, by (2.7),

$$\mathcal{A}(f,f) + \|g\|^2 = \|A^{1/2} f\|^2 + \|g\|^2 < \infty \tag{3.1}$$

or, equivalently,

$$f \in L_2^1(R^3), \quad g \in \mathcal{H}. \tag{3.2}$$

It follows that the function u(t,X) defined by (1.14) is the unique solution wFE and

$$u \in C^1(R,\mathcal{H}) \cap C(R,L_2^1(R^3)), \tag{3.3}$$

(Here $C^k(I,\mathcal{H})$ is the set of k times continuously differentiable $\mathcal{H}$-valued function on I.)

The acoustic fields studied below are defined by (1.14) where f, g satisfy (3.2). It will be convenient to specialize further by writing u in the form (1.19)-(1.21). This is correct provided f and g are real valued, (3.2) holds and $g \in D(A^{-1/2})$. Of course, $D(A^{-1/2}) \neq \mathcal{H}$, in general. However, $D(A^{-1/2})$ is dense in $\mathcal{H}$, by the spectral theorem. This fact can often be used to extend results derived from (1.19)-(1.21) to general solutions wFE.

Chapter 3
Spectral Analysis of Sound Propagation in Stratified Fluids

The purpose of this chapter is to construct families of normal mode functions and to derive corresponding normal mode expansions for the acoustic propagator A. The basic physical concepts and results are summarized in §1. A more detailed formulation and complete proofs of all the results are developed in the remaining sections. This rather long chapter provides the analytical foundation for the solutions of the propagation problems that are developed in Chapters 4 and 5.

§1. SUMMARY

Throughout this chapter it is assumed that $\rho(y)$ and $c(y)$ satisfy the boundedness conditions (2.1.4) and the four conditions

$$(1.1) \qquad \pm\int_0^{\pm\infty} |\rho(y) - \rho(\pm\infty)|\, dy < \infty, \quad \pm\int_0^{\pm\infty} |c(y) - c(\pm\infty)|\, dy < \infty,$$

where $\rho(+\infty)$, $\rho(-\infty)$, $c(+\infty)$ and $c(-\infty)$ are suitable constants. Under these conditions the spectrum of A fills the interval $[0,\infty)$ and is continuous with no embedded eigenvalues. These facts, which are verified below, imply that the normal mode functions of A are generalized eigenfunctions; that is, solutions ψ of the differential equation

$$(1.2) \qquad A\psi \equiv -c^2(y)\, \rho(y)\, \nabla \cdot \left(\frac{1}{\rho(y)} \nabla\psi(x,y)\right) = \lambda\, \psi(x,y)$$

that are not in $\mathcal{H}$. Solution of (1.2) by separation of variables leads to solutions of the form

$$(1.3) \qquad \psi(x,y) = e^{ip\cdot x}\, \psi(y), \quad p = (p_1,p_2) \in R^2,$$

where $p \cdot x = p_1 x_1 + p_2 x_2$ and $\psi(y)$ is a solution of the equation

$$-c^2(y)\left[\rho(y)\frac{d}{dy}\left(\frac{1}{\rho(y)}\frac{d\psi}{dy}\right) - |p|^2\psi\right] = \lambda\psi \tag{1.4}$$

with $|p|^2 = p_1^2 + p_2^2$.

The operator A_μ defined by

$$A_\mu\psi = -c^2(y)\left[\rho(y)\frac{d}{dy}\left(\frac{1}{\rho(y)}\frac{d\psi}{dy}\right) - \mu^2\psi\right] \tag{1.5}$$

will be called the reduced acoustic propagator. It is a singular Sturm-Liouville operator on $-\infty < y < \infty$. Its proper and improper eigenfunctions will be used to generate the normal mode functions of A.

Note that (2.1.4) and (1.1) imply that

$$\rho_m \le \rho(\pm\infty) \le \rho_M, \quad c_m \le c(\pm\infty) \le c_M. \tag{1.6}$$

Moreover, in the presence of (2.1.4), conditions (1.1) are restrictions on the behavior of $\rho(y)$ and $c(y)$ near $y = \pm\infty$ only. Thus it is natural to compare solutions of $A_\mu\phi = \lambda\phi$ with solutions of its limiting forms for $y \to \pm\infty$:

$$\frac{d^2\phi}{dy^2} + (\lambda\, c^{-2}(\pm\infty) - \mu^2)\phi = 0. \tag{1.7}$$

In what follows it will be convenient to assume that

$$c(\infty) \le c(-\infty). \tag{1.8}$$

This condition does not limit the generality since it can always be achieved by proper choice of the depth coordinate y.

The general solution of (1.7) for $\lambda \ne c^2(\pm\infty)\mu^2$ can be written

$$\left\{\begin{array}{l} \phi = c_1\, e^{iq_\pm(\mu,\lambda)y} + c_2\, e^{-iq_\pm(\mu,\lambda)y}, \text{ or} \\ \\ \phi = c_1\, e^{q'_\pm(\mu,\lambda)y} + c_2\, e^{-q'_\pm(\mu,\lambda)y} \end{array}\right. \tag{1.9}$$

where

$$\left.\begin{array}{l} q_\pm(\mu,\lambda) = (\lambda c^{-2}(\pm\infty) - \mu^2)^{1/2} > 0 \\ \\ q'_\pm(\mu,\lambda) = iq_\pm(\mu,\lambda) \end{array}\right\} \text{ for } \lambda > c^2(\pm\infty)\mu^2 \tag{1.10}$$

and

$$(1.11) \qquad \left.\begin{cases} q_\pm'(\mu,\lambda) = (\mu^2 - \lambda c^{-2}(\pm\infty))^{1/2} > 0 \\ \\ q_\pm(\mu,\lambda) = -iq_\pm'(\mu,\lambda) \end{cases}\right\} \text{ for } \lambda < c^2(\pm\infty)\mu^2$$

In particular, the solutions are oscillatory when $\lambda > c^2(\pm\infty)\mu^2$ and non-oscillatory when $\lambda < c^2(\pm\infty)\mu^2$. The first step in the analysis of A_μ is to construct special solutions of $A_\mu\phi = \lambda\phi$ that have the asymptotic behaviors

$$(1.12) \qquad \begin{cases} \left.\begin{aligned} \phi_1(y,\mu,\lambda) &= e^{q_+'(\mu,\lambda)y}[1 + o(1)], \; y \to +\infty \\ \phi_2(y,\mu,\lambda) &= e^{-q_+'(\mu,\lambda)y}[1 + o(1)], \; y \to +\infty \end{aligned}\right\} \lambda \neq c^2(\infty)\mu^2, \\ \left.\begin{aligned} \phi_3(y,\mu,\lambda) &= e^{q_-'(\mu,\lambda)y}[1 + o(1)], \; y \to -\infty \\ \phi_4(y,\mu,\lambda) &= e^{-q_-'(\mu,\lambda)y}[1 + o(1)], \; y \to -\infty \end{aligned}\right\} \lambda \neq c^2(-\infty)\mu^2. \end{cases}$$

It follows from an asymptotic calculation of the Wronskians that ϕ_1 and ϕ_2 are a solution basis for $A_\mu\phi = \lambda\phi$ when $\lambda \neq c^2(\infty)\mu^2$, while ϕ_3 and ϕ_4 are a basis when $\lambda \neq c^2(-\infty)\mu^2$.

The nature of the spectrum and eigenfunctions of A_μ can be inferred from (1.12). It follows that if $\lambda < c^2(\infty)\mu^2$ then $A_\mu\phi = \lambda\phi$ has bounded solutions only if ϕ_2 and ϕ_3 are linearly dependent. Thus

$$(1.13) \qquad F(\mu,\lambda) = \rho^{-1} W(\phi_2,\phi_3) = 0$$

is an equation for the eigenvalues of A_μ, where W denotes the Wronskian and $\rho^{-1}W$ is independent of y. The corresponding solutions

$$(1.14) \qquad \psi_k(y,\mu) = a_k(\mu)\,\phi_2(y,\mu,\lambda_k(\mu)) = a_k'(\mu)\,\phi_3(y,\mu,\lambda_k(\mu)),$$

where $\lambda = \lambda_k(\mu)$ is a root of (1.13), are square integrable on R and hence are eigenfunctions of A_μ. Moreover, A_μ can have no point eigenvalues $\lambda > c^2(\infty)\mu^2$, by (1.12). Thus $\sigma_0(A_\mu)$, the point spectrum of A_μ, lies in the interval $[c_m^2\mu^2, c^2(\infty)\mu^2]$. Criteria for $\sigma_0(A_\mu)$ to be empty, finite or countably infinite are given in §4 below.

It will be shown that the continuous spectrum of A_μ is $[c^2(\infty)\mu^2,\infty)$ and corresponding generalized eigenfunctions will be determined from (1.12).

For $c^2(\infty)\mu^2 < \lambda < c^2(-\infty)\mu^2$ there is a single family of generalized eigenfunctions of the form

$$\psi_0(y,\mu,\lambda) = a_0(\mu,\lambda)\,\phi_3(y,\mu,\lambda). \tag{1.15}$$

For $\lambda > c^2(-\infty)\mu^2$ there are two families defined by

$$\begin{cases} \psi_+(y,\mu,\lambda) = a_+(\mu,\lambda)\,\phi_4(y,\mu,\lambda) \\ \psi_-(y,\mu,\lambda) = a_-(\mu,\lambda)\,\phi_1(y,\mu,\lambda) \end{cases} \tag{1.16}$$

It will be shown that these functions have the following asymptotic forms.

$$\psi_k(y,\mu) \sim \begin{cases} c_k^+(\mu)\, e^{-q_+'(\mu,\lambda_k(\mu))y}, & y \to +\infty, \\ c_k^-(\mu)\, e^{q_-'(\mu,\lambda_k(\mu))y}, & y \to -\infty, \end{cases} \tag{1.17}$$

$$\psi_0(y,\mu,\lambda) \sim c_0(\mu,\lambda) \begin{cases} e^{-iq_+(\mu,\lambda)y} + R_0(\mu,\lambda)\, e^{iq_+(\mu,\lambda)y}, & y \to +\infty, \\ T_0(\mu,\lambda)\, e^{q_-'(\mu,\lambda)y}, & y \to -\infty, \end{cases} \tag{1.18}$$

$$\psi_+(y,\mu,\lambda) \sim c_+(\mu,\lambda) \begin{cases} e^{-iq_+(\mu,\lambda)y} + R_+(\mu,\lambda)\, e^{iq_+(\mu,\lambda)y}, & y \to +\infty, \\ T_+(\mu,\lambda)\, e^{-iq_-(\mu,\lambda)y}, & y \to -\infty, \end{cases} \tag{1.19}$$

$$\psi_-(y,\mu,\lambda) \sim c_-(\mu,\lambda) \begin{cases} T_-(\mu,\lambda)\, e^{iq_+(\mu,\lambda)y}, & y \to +\infty, \\ e^{iq_-(\mu,\lambda)y} + R_-(\mu,\lambda)\, e^{-iq_-(\mu,\lambda)y}, & y \to -\infty. \end{cases} \tag{1.20}$$

Here $a_k(\mu)$, $a_k'(\mu)$, $a_0(\mu,\lambda)$, $a_\pm(\mu,\lambda)$, $c_k^\pm(\mu)$, $c_0(\mu,\lambda)$, $c_\pm(\mu,\lambda)$, $R_0(\mu,\lambda)$, $R_\pm(\mu,\lambda)$, $T_0(\mu,\lambda)$ and $T_\pm(\mu,\lambda)$ are functions of μ and λ that will be calculated below.

Families of normal mode functions for A may be constructed from those for $A_{|p|}$ by the rule (1.3). The following notation will be used.

$$\psi_\pm(x,y,p,\lambda) = (2\pi)^{-1} e^{ip\cdot x} \psi_\pm(y,|p|,\lambda), \ (p,\lambda) \in \Omega, \tag{1.21}$$

$$\psi_0(x,y,p,\lambda) = (2\pi)^{-1} e^{ip\cdot x} \psi_0(y,|p|,\lambda), \ (p,\lambda) \in \Omega_0, \tag{1.22}$$

$$\psi_k(x,y,p) = (2\pi)^{-1} e^{ip\cdot x} \psi_k(y,|p|), \quad p \in \Omega_k, \ k \geq 1, \tag{1.23}$$

The parameter domains Ω, Ω_0, Ω_k are defined by

$$\Omega = \{(p,\lambda) \in R^3 \mid c^2(-\infty) \ |p|^2 < \lambda\}, \tag{1.24}$$

$$\Omega_0 = \{(p,\lambda) \in R^3 \mid c^2(\infty) \ |p|^2 < \lambda < c^2(-\infty) \ |p|^2\} \tag{1.25}$$

$$\Omega_k = \{p \in R^2 \mid |p| \in \mathcal{O}_k\}, \ k \geq 1, \tag{1.26}$$

where $\mathcal{O}_k$ is the set of $\mu > 0$ for which A_μ has a kth eigenvalue.

The three families have different wave-theoretic interpretations that are characterized by their asymptotic behaviors. Thus for $(p,\lambda) \in \Omega$ one has

$$\psi_+(x,y,p,\lambda) \sim \frac{c_+(|p|,\lambda)}{2\pi} \begin{cases} e^{i(p\cdot x-q_+y)} + R_+ e^{i(p\cdot x+q_+y)}, & y \to +\infty, \\ T_+ e^{i(p\cdot x-q_-y)}, & y \to -\infty, \end{cases} \tag{1.27}$$

$$\psi_-(x,y,p,\lambda) \sim \frac{c_-(|p|,\lambda)}{2\pi} \begin{cases} T_- e^{i(p\cdot x+q_+y)}, & y \to +\infty, \\ e^{i(p\cdot x+q_-y)} + R_- e^{i(p\cdot x-q_-y)}, & y \to -\infty, \end{cases} \tag{1.28}$$

where $q_\pm = q_\pm(|p|,\lambda)$, $R_+ = R_+(|p|,\lambda)$, etc. Hence $\psi_+(x,y,p,\lambda)$ behaves for $y \to +\infty$ like an incident plane wave with propagation vector $\vec{k}_i = (p,-q_+)$ plus a specularly reflected wave with propagation vector $\vec{k}_r = (p,q_+)$, while for $y \to -\infty$ it behaves like a pure transmitted plane wave with propagation vector $\vec{k}_t = (p,-q_-)$. The incident and transmitted plane waves can be shown to satisfy Snell's law $n(\infty) \sin \theta(\infty) = n(-\infty) \sin \theta(-\infty)$ where $\theta(\infty)$ and $\theta(-\infty)$ are the angles between the y-axis and $\vec{k}_i$ and $\vec{k}_t$, respectively, and $n(\pm\infty) = c^{-1}(\pm\infty)$. $\psi_-(x,y,p,\lambda)$ has a similar interpretation.

For $(p,\lambda) \in \Omega_0$ one has

$$(1.29) \qquad \psi_0(x,y,p,\lambda) \sim \frac{c_0(|p|,\lambda)}{2\pi} \begin{cases} e^{i(p\cdot x - q_+ y)} + R_0\, e^{i(p\cdot x + q_+ y)}, & y \to +\infty, \\ T_0\, e^{ip\cdot x}\, e^{q'_- y}, & y \to -\infty. \end{cases}$$

Hence for $y \to +\infty$ $\psi_0(x,y,p,\lambda)$ behaves like an incident plane wave plus a specularly reflected wave while for $y \to -\infty$ it is exponentially damped. This is analogous to the phenomenon of total reflection of a plane wave in a homogeneous medium of refractive index $n(\infty) = c^{-1}(\infty)$ at an interface with a medium of index $n(-\infty) = c^{-1}(-\infty) < n(\infty)$. Indeed, the condition $\lambda < c^2(-\infty)\,|p|^2$ is equivalent to the condition for total reflection: $n(\infty) \sin\theta(\infty) > n(-\infty)$.

For $p \in \Omega_k$, $k \geq 1$, one has

$$(1.30) \qquad \psi_k(x,y,p) \sim \begin{cases} \dfrac{c_k^+(|p|)}{2\pi}\, e^{ip\cdot x}\, e^{-q'_+ y}, & y \to +\infty, \\ \dfrac{c_k^-(|p|)}{2\pi}\, e^{ip\cdot x}\, e^{q'_- y}, & y \to -\infty. \end{cases}$$

Hence the functions $\psi_k(x,y,p)$ can be interpreted as guided waves that are trapped by total reflection in the acoustic duct where $c(y) < c(\pm\infty)$. They propagate in the direction $\vec{k} = (p,0)$ parallel to the duct and decrease exponentially with distance from it.

The coefficients $R_\pm$, R_0 and $T_\pm$, T_0 in (1.27), (1.28), (1.29) may be interpreted as reflection and transmission coefficients, respectively, for the scattering of plane waves by the stratified fluid. They will be shown to satisfy the conservation laws

$$(1.31) \qquad \frac{q_\pm}{\rho(\pm\infty)}\,|R_\pm|^2 + \frac{q_\mp}{\rho(\mp\infty)}\,|T_\pm|^2 = \frac{q_\pm}{\rho(\pm\infty)}, \qquad |R_0| = 1.$$

The three families ψ_+, ψ_- and ψ_0 represent, collectively, the response of the stratified fluid to incident plane waves $\exp\{i(p\cdot x - qy)\}$, $(p,q) \in R^3$. To see this consider the mappings

$$(1.32) \qquad \begin{cases} (p,q) = X_+(p,\lambda) = (p, q_+(|p|,\lambda)), & (p,\lambda) \in \Omega, \\ (p,q) = X_0(p,\lambda) = (p, q_+(|p|,\lambda)), & (p,\lambda) \in \Omega_0, \\ (p,q) = X_-(p,\lambda) = (p, -q_-(|p|,\lambda)), & (p,\lambda) \in \Omega. \end{cases}$$

X_+ is an analytic transformation of Ω onto the cone

$$C_+ = \{(p,q) \mid q > a\,|p|\} \tag{1.33}$$

where

$$a = ((c(-\infty)/c(\infty))^2 - 1)^{1/2} \geq 0. \tag{1.34}$$

Similarly, X_0 is an analytic transformation of Ω_0 onto the cone

$$C_0 = \{(p,q) \mid 0 < q < a\,|p|\} \tag{1.35}$$

and X_- is an analytic transformation of Ω onto the cone

$$C_- = \{(p,q) \mid q < 0\}. \tag{1.36}$$

Thus, the asymptotic forms of $\psi_\pm$ and ψ_0 for $y \to \pm\infty$ show that $\psi_+(x,y,p,\lambda)$ with $(p,\lambda) \in \Omega$ is the response of the fluid to a plane wave $\exp\{i(p\cdot x - qy)\}$ with $(p,q) \in C_+$, $\psi_0(x,y,p,\lambda)$ is the response to a plane wave with $(p,q) \in C_0$ and $\psi_-(x,y,p,\lambda)$ is the response to a plane wave with $(p,q) \in C_-$. Note that

$$R^3 = C_+ \cup C_0 \cup C_- \cup N \tag{1.37}$$

where $N = \partial C_+ \cup \partial C_-$, $\partial C_+ = \{(p,q) : q = a\,|p|\}$ and $\partial C_- = \{(p,q) : q = 0\}$. Physically, vectors $(p,-q)$ with $(p,q) \in \partial C_+$ separate the plane waves incident from $y = +\infty$ that have a transmitted plane wave at $y = -\infty$ from those that are totally reflected. Vectors $(p,0) \in \partial C_-$ correspond to plane waves that are incident parallel to the stratification. The momenta $(p,q) \in N$ form a Lebesgue null set in the momentum space R^3 and are not needed in the theory developed here.

The interpretation of ψ_+, ψ_- and ψ_0 given above suggests the introduction of a composite normal mode function

$$\phi_+(x,y,p,q) = (2\pi)^{-1}\, e^{ip\cdot x}\, \phi_+(y,p,q),\quad (p,q) \in R^3, \tag{1.38}$$

where

$$(1.39) \quad \phi_+(y,p,q) = \begin{cases} (2q)^{1/2}\, c(\infty)\ \psi_+(y,|p|,\lambda), & (p,\lambda) = X_+^{-1}(p,q),\ (p,q) \in C_+, \\ (2q)^{1/2}\, c(\infty)\ \psi_0(y,|p|,\lambda), & (p,\lambda) = X_0^{-1}(p,q),\ (p,q) \in C_0, \\ (2|q|)^{1/2}\, c(-\infty)\ \psi_-(y,|p|,\lambda), & (p,\lambda) = X_-^{-1}(p,q),\ (p,q) \in C_-, \\ 0 & , (p,q) \in N. \end{cases}$$

The normalizing factors $(2q)^{1/2}\, c(\infty)$ and $(2|q|)^{1/2}\, c(-\infty)$ are the square roots of the Jacobians of X_+^{-1}, X_0^{-1} and X_-^{-1}. The function $\phi_+(x,y,p,q)$ is a solution of the differential equation

$$(1.40) \qquad A\ \phi_+(\cdot,p,q) = \lambda(p,q)\ \phi_+(\cdot,p,q)$$

where

$$(1.41) \qquad \lambda(p,q) = \begin{cases} c^2(\infty)(|p|^2 + q^2), & (p,q) \in C_+ \cup C_0, \\ c^2(-\infty)(|p|^2 + q^2), & (p,q) \in C_-, \\ 0 & , (p,q) \in N. \end{cases}$$

Its asymptotic behavior is described by

$$(1.42) \qquad \phi_+(x,y,p,q) \sim c(p,q) \begin{cases} e^{i(p\cdot x-qy)} + R_+\ e^{i(p\cdot x+qy)}, & (p,q) \in C_+, \\ e^{i(p\cdot x-qy)} + R_0\ e^{i(p\cdot x+qy)}, & (p,q) \in C_0, \\ T_-\ e^{i(p\cdot x+q_+(|p|,\lambda)y)} & , (p,q) \in C_-, \end{cases}$$

for $y \to +\infty$ and

$$(1.43) \qquad \phi_+(x,y,p,q) \sim c(p,q) \begin{cases} T_+\ e^{i(p\cdot x-q_-(|p|,\lambda)y)}, & (p,q) \in C_+, \\ T_0\ e^{ip\cdot x}\ e^{q'_-(|p|,\lambda)y}, & (p,q) \in C_0, \\ e^{i(p\cdot x-qy)} + R_-\ e^{i(p\cdot x+qy)}, & (p,q) \in C_- \end{cases}$$

for $y \to -\infty$. In §9 it is shown that one may take

$$(1.44)\qquad c(p,q) = \begin{cases} (2\pi)^{-3/2}\, c(\infty)\, \rho^{1/2}(\infty), & (p,q) \in C_+ \cup C_0, \\ (2\pi)^{-3/2}\, c(-\infty)\, \rho^{1/2}(-\infty), & (p,q) \in C_-. \end{cases}$$

Another family of normal mode functions for A is defined by

$$(1.45)\qquad \phi_-(x,y,p,q) = \overline{\phi_+(x,y,-p,q)}, \quad (p,q) \in R^3.$$

It is clear that $A\,\phi_- = \lambda(p,q)\,\phi_-$ and

$$(1.46)\qquad \phi_-(x,y,p,q) = (2\pi)^{-1}\, e^{ip\cdot x}\, \phi_-(y,p,q)$$

where

$$(1.47)\qquad \phi_-(y,p,q) = \overline{\phi_+(y,p,q)}.$$

The asymptotic behavior of ϕ_- for $y \to \pm\infty$ may be derived from (1.42), (1.43), (1.44). It is given by

$$(1.48)\qquad \phi_-(x,y,p,q) \sim c(p,q) \begin{cases} e^{i(p\cdot x+qy)} + \overline{R}_+\, e^{i(p\cdot x-qy)}, & (p,q) \in C_+, \\ e^{i(p\cdot x+qy)} + \overline{R}_0\, e^{i(p\cdot x-qy)}, & (p,q) \in C_0, \\ \overline{T}_-\, e^{i(p\cdot x-q_+(|p|,\lambda)y)}, & (p,q) \in C_-, \end{cases}$$

for $y \to +\infty$ and

$$(1.49)\qquad \phi_-(x,y,p,q) \sim c(p,q) \begin{cases} \overline{T}_+\, e^{i(p\cdot x+q_-(|p|,\lambda)y)}, & (p,q) \in C_+, \\ \overline{T}_0\, e^{ip\cdot x}\, e^{q_-'(|p|,\lambda)y}, & (p,q) \in C_0, \\ e^{i(p\cdot x+qy)} + \overline{R}_-\, e^{i(p\cdot x-qy)}, & (p,q) \in C_-, \end{cases}$$

for $y \to -\infty$. These relations clearly imply that $\phi_-(x,y,p,q)$ is not simply a multiple of $\phi_+(x,y,p,q)$. By contrast the guided mode functions have the symmetry property

(1.50) $$\psi_k(x,y,p) = \overline{\psi_k(x,y,-p)}, \quad k \geq 1,$$

because the functions $\psi_k(y,|p|)$ are real-valued and depend on p only through $|p|$.

The completeness of the family $\{\phi_-,\psi_1,\psi_2,\ldots\}$ is derived from that of $\{\phi_+,\psi_1,\psi_2,\ldots\}$ in §9. The existence of the two families ϕ_+ and ϕ_- is a consequence of the invariance of the wave equation under time reversal. In Chapter 4 the family ϕ_- is used to construct asymptotic solutions for $t \to +\infty$ of the propagation problem.

The normal mode expansions which are the main results of this chapter will now be formulated. For clarity, the case in which A has no guided modes is described first. The general case is described at the end of the section.

The normal mode expansions for A are in essence Hilbert space expansions; that is, they converge in a mean square, or Hilbert space, sense. The space $\mathcal{H}$ in which A acts was defined by (2.1.13). The Lebesgue spaces $L_2(Q)$ of square integrable functions on $Q \subset R^n$ are also needed. $L_2(Q)$ is a Hilbert space with scalar product

(1.51) $$(u,v) = \int_Q \overline{u(P)}\, v(P)\, dP$$

where $dP = dp_1 \ldots dp_n$ is the volume element in R^n.

If A has no guided modes then the family $\{\phi_+(X,P) : P \in R^3\}$ is a complete family of normal mode functions for A. This statement means that every $h \in \mathcal{H}$ has a unique generalized Fourier transform $\hat{h}_+ \in L_2(R^3)$ such that

(1.52) $$\hat{h}_+(P) = L_2(R^3)\text{-lim} \int_{R^3} \overline{\phi_+(X,P)}\, h(X)\, c^{-2}(y)\, \rho^{-1}(y)\, dX,$$

(1.53) $$\|h\|_{\mathcal{H}} = \|\hat{h}_+\|_{L_2(R^3)}$$

and

(1.54) $$h(X) = \mathcal{H}\text{-lim} \int_{R^3} \phi_+(X,P)\, \hat{h}_+(P)\, dP.$$

Equation (1.53) generalizes the Parseval relation of Fourier analysis. Equation (1.52) is a condensed notation for the assertion that if $\{K_m\}$ is <u>any</u> nested sequence of compact sets in R^3 such that $\cup K_m = R^3$, and if

$$(1.55)\qquad h_m(X) = \begin{cases} h(X), & X \in K_m, \\ 0, & X \in R^3 - K_m, \end{cases}$$

then the integrals

$$(1.56)\qquad \int_{R^3} \overline{\phi_+(X,P)}\, h_m(X)\, c^{-2}(y)\, \rho^{-1}(y)dX = \int_{K_m} \overline{\phi_+(X,P)}\, h(X)\, c^{-2}(y)\rho^{-1}(y)dX$$

are finite for every $P \in R^3$ and define functions $\hat{h}_{m+} \in L_2(R^3)$ such that the sequence $\{h_{m+}\}$ is convergent in $L_2(R^3)$. $\hat{h}_+$ is by definition the limit of this sequence.

Equation (1.54) has the analogous interpretation. Thus it represents an arbitrary $h \in \mathcal{H}$ as a superposition of normal mode functions. (1.54) is useful because it provides a spectral representation of A. This means that if h is in the domain of A, so that $Ah \in \mathcal{H}$, then $\lambda(P)\, \hat{h}_+(P) \in L_2(R^3)$ and

$$(1.57)\qquad (Ah)\hat{}_+(P) = \lambda(P)\, \hat{h}_+(P)$$

where $\lambda(P)$ is defined by (1.41). This property is used in Chapter 4 to solve the wave equation for A.

In what follows the $\mathcal{H}$-lim notation is sometimes suppressed for the sake of simplicity. However, integrals such as (1.52) and (1.54) are always to be interpreted as Hilbert space limits. Of course, for special choices of h it may happen that the integrals also converge pointwise. However, even in these cases it is the mean square convergence that is the more relevant to problems of wave propagation.

To describe the general case where A has guided modes let $N_0 - 1$ denote the number of distinct guided normal mode functions ψ_k for A where $2 \le N_0 \le +\infty$. Then (1.52) still holds for every $h \in \mathcal{H}$ and in addition for every k such that $1 \le k < N_0$ there exists a unique function $\tilde{h}_k \in L_2(\Omega_k)$ such that

$$(1.58)\qquad \tilde{h}_k(p) = L_2(\Omega_k)\text{-lim} \int_{R^3} \overline{\psi_k(X,p)}\, h(X)\, c^{-2}(y)\, \rho^{-1}(y)dX.$$

Moreover, the Parseval relation (1.53) is replaced by

$$(1.59)\qquad \|h\|^2_{\mathcal{H}} = \|\hat{h}_+\|^2_{L_2(R^3)} + \sum_{k=1}^{N_0-1} \|\tilde{h}_k\|^2_{L_2(\Omega_k)}$$

and the normal mode expansions asserts that the limits

(1.60)
$$h_f(X) = \mathcal{H}\text{-lim} \int_{R^3} \phi_+(X,P)\, \hat{h}_+(P)\,dP$$

and

(1.61)
$$h_k(X) = \mathcal{H}\text{-lim} \int_{\Omega_k} \psi_k(X,p)\, \tilde{h}_k(p)\,dp, \quad 1 \le k < N_0,$$

exist and

(1.62)
$$h = h_f + \sum_{k=1}^{N_0-1} h_k$$

where if $N_0 = +\infty$ then the last sum is convergent in $\mathcal{H}$. The representation (1.52), (1.58)-(1.62) is a spectral representation for A in the sense that if h is in the domain of A then (1.57) holds and in addition

(1.63)
$$(Ah)_k^{\sim}(p) = \lambda_k(|p|)\, \tilde{h}_k(p) \quad \text{for } 1 \le k < N_0$$

where $A\psi_k(X,p) = \lambda_k(|p|)\, \psi_k(X,p)$.

The functions $\{\phi_-,\psi_1,\psi_2,\cdots\}$ form a second complete set of normal mode functions for A. The normal mode expansion for this family is defined by (1.52), (1.58)-(1.62) with ϕ_+ and $\hat{h}_+$ replaced by ϕ_- and $\hat{h}_-$. The functions $\{\psi_+,\psi_-,\psi_0,\psi_1,\psi_2,\cdots\}$ form still another complete set. The normal mode expansion for this family is derived in §9.

§2. THE REDUCED PROPAGATOR A_μ

The Sturm-Liouville operator A_μ defined by (1.5) has a selfadjoint realization, also to be denoted by A_μ, in the Hilbert space

(2.1)
$$\mathcal{H}(R) = L_2(R, c^{-2}(y)\ \rho^{-1}(y)dy).$$

The domain of A_μ is the set

(2.2)
$$D(A_\mu) = L_2^1(R) \cap \left\{\psi : \frac{d}{dy}\left(\rho^{-1}(y)\ \frac{d\psi}{dy}\right) \in L_2(R)\right\}.$$

The properties

(2.3)
$$A_\mu = A_\mu^* \ge c_m^2\mu^2,$$

where A_μ^* is the adjoint of A_μ in $\mathcal{H}(R)$, can be verified by showing that A_μ is the operator in $\mathcal{H}(R)$ associated with the sesquilinear form A_μ in $\mathcal{H}(R)$

defined by

$$D(A_\mu) = L_2^1(R) \subset \mathcal{H}(R) \tag{2.4}$$

and

$$A_\mu(\phi,\psi) = \int_R \left\{ \frac{d\overline{\phi}}{dy} \frac{d\psi}{dy} + \mu^2 \overline{\phi}\psi \right\} \rho^{-1}(y)dy. \tag{2.5}$$

The spectral analysis of A will be derived below from that of A_μ. The main steps of the analysis are the following. First, conditions (2.1.4) and (1.1) are used to construct the special solutions of $A_\mu \phi = \lambda\phi$ defined by conditions (1.12). Second, these solutions are used to construct an eigenfunction expansion for A_μ. The construction is based on the Weyl-Kodaira-Titchmarsh theory of singular Sturm-Liouville operators. Finally, the expansion for $A_{|p|}$ and Fourier analysis in the variables x_1, x_2 are used to construct a spectral representation for A. This method has been applied to the special cases of the Pekeris and Epstein profiles [8,26] where explicit representations of the solutions of $A_\mu \phi = \lambda\phi$ by means of elementary functions are available. Thus the main technical advance in the work presented in this chapter is the construction, for the class of density and sound speed profiles defined by (2.1.4) and (1.1), of solutions of $A_\mu \phi = \lambda\phi$ that have prescribed asymptotic behavior for $y \to \pm\infty$ and sufficient regularity in the parameters λ and μ to permit application of the methods of [8,26].

§3. SOLUTIONS OF THE EQUATION $A_\mu \phi = \zeta\phi$

The special solutions $\phi_j(y,\mu,\lambda)$ ($j = 1,2,3,4$) described in §1 are constructed in this section. Analytic continuations of these functions to complex values of λ are used in §6 for the calculation of the spectral family of A_μ. Hence the more general case of solutions of $A_\mu \phi = \zeta\phi$ with $\zeta \in C$ will be treated.

The equation $A_\mu \phi = \zeta\phi$ cannot have solutions in the classical sense unless $c(y)$ and $\rho(y)$ are continuous and continuously differentiable, respectively. A suitable class of solutions is described by the following definition in which AC(I) denotes the set of all functions that are absolutely continuous with respect to Lebesgue measure in the interval $I = (a,b) \subset R$.

<u>Definition</u>. A function $\phi : I = (a,b) \to C$ is said to be a solution of

$$A_\mu \phi(y) \equiv -c^2(y)\{\rho(y)(\rho^{-1}(y)\ \phi'(y))' - \mu^2\phi(y)\} = \zeta\phi(y) \tag{3.1}$$

in the interval I (where $\phi' = d\phi/dy$) if and only if

$$\phi \in AC(I), \ \rho^{-1}\phi' \in AC(I) \tag{3.2}$$

and (3.1) holds for almost all $y \in I$.

The following notation will be used in the definition and construction of the special solutions $\phi_j(y,\mu,\zeta)$. For each $\kappa \geq 0$

$$\begin{aligned} L(\kappa) &= \{\zeta \mid \text{Re}\ \zeta < \kappa^2\}, \\ R(\kappa) &= \{\zeta \mid \text{Re}\ \zeta > \kappa^2\}, \\ R^{\pm}(\kappa) &= R(\kappa) \cap \{\zeta \mid \pm\text{Im}\ \zeta \geq 0\}. \end{aligned} \tag{3.3}$$

The definitions (1.10), (1.11) will be extended as follows.

$$\left.\begin{aligned} q_{\pm}(\mu,\zeta) &= (\zeta\, c^{-2}(\pm\infty) - \mu^2)^{1/2} \\ -\pi/4 &< \arg q_{\pm}(\mu,\zeta) < \pi/4 \\ q'_{\pm}(\mu,\zeta) &= i\, q_{\pm}(\mu,\zeta) \end{aligned}\right\} \ \zeta \in R(c(\pm\infty)\mu) \tag{3.4}$$

and

$$\left.\begin{aligned} q'_{\pm}(\mu,\zeta) &= (\mu^2 - \zeta\, c^{-2}(\pm\infty))^{1/2} \\ -\pi/4 &< \arg q'_{\pm}(\mu,\zeta) < \pi/4 \\ q_{\pm}(\mu,\zeta) &= -i\, q'_{\pm}(\mu,\zeta). \end{aligned}\right\} \ \zeta \in L(c(\pm\infty)\mu) \tag{3.5}$$

The results of this section will now be formulated.

Theorem 3.1. Under hypotheses (2.1.4), (1.1) on $\rho(y)$, $c(y)$ there exist functions

$$\phi_j : R \times R_+ \times (L(c(\infty)\mu) \cup R(c(\infty)\mu)) \to C, \ j = 1,2, \tag{3.6}$$

(where $R_+ = \{\mu \mid \mu \geq 0\}$) such that for every fixed $(\mu,\zeta) \in R_+ \times (L(c(\infty)\mu) \cup R(c(\infty)\mu))$, $\phi_j(y,\mu,\zeta)$ is a solution of (3.1) for $y \in R$ and $j = 1,2$ and

$$(3.7)\quad \left.\begin{aligned} \phi_1(y,\mu,\zeta) &= \exp\{q_+'(\mu,\zeta)y\}[1+o(1)] \\ \rho^{-1}(y)\ \phi_1'(y,\mu,\zeta) &= \rho^{-1}(\infty)\ q_+'(\mu,\zeta)\ \exp\{q_+'(\mu,\zeta)y\}[1+o(1)] \end{aligned}\right\}\ y \to +\infty,$$

and

$$(3.8)\quad \left.\begin{aligned} \phi_2(y,\mu,\zeta) &= \exp\{-q_+'(\mu,\zeta)y\}[1+o(1)] \\ \rho^{-1}(y)\ \phi_2'(y,\mu,\zeta) &= -\rho^{-1}(\infty)\ q_+'(\mu,\zeta)\ \exp\{-q_+'(\mu,\zeta)y\}[1+o(1)] \end{aligned}\right\}\ y \to +\infty.$$

Similarly, there exist functions

$$(3.9)\quad \phi_j : R \times R_+ \times (L(c(-\infty)\mu) \cup R(c(-\infty)\mu)) \to C,\ j = 3,4,$$

such that for every fixed $(\mu,\zeta) \in R_+ \times (L(c(-\infty)\mu) \cup R(c(-\infty)\mu)$, $\phi_j(y,\mu,\zeta)$ is a solution of (3.1) for $y \in R$ and $j = 3,4$ and

$$(3.10)\quad \left.\begin{aligned} \phi_3(y,\mu,\zeta) &= \exp\{q_-'(\mu,\zeta)y\}[1+o(1)] \\ \rho^{-1}(y)\ \phi_3'(y,\mu,\zeta) &= \rho^{-1}(-\infty)\ q_-'(\mu,\zeta)\ \exp\{q_-'(\mu,\zeta)y\}[1+o(1)] \end{aligned}\right\}\ y \to -\infty,$$

and

$$(3.11)\quad \left.\begin{aligned} \phi_4(y,\mu,\zeta) &= \exp\{-q_-'(\mu,\zeta)y\}[1+o(1)] \\ \rho^{-1}(y)\phi_4'(y,\mu,\zeta) &= -\rho^{-1}(-\infty)q_-'(\mu,\zeta)\ \exp\{-q_-'(\mu,\zeta)y\}[1+o(1)] \end{aligned}\right\}\ y \to -\infty.$$

The following three corollaries describe the dependence of the solutions $\phi_j(y,\mu,\zeta)$ on the parameters μ and ζ.

Corollary 3.2. The functions $\phi_j(y,\mu,\zeta)$ satisfy

$$(3.12)\quad \phi_j,\ \rho^{-1}\phi_j' \in C\left(R \times \bigcup_{\mu\geq 0} \{(\mu,\zeta)\ |\ \zeta \in L(c(\infty)\mu)\}\right)$$

for j = 1, 2 and

$$(3.13)\quad \phi_j,\ \rho^{-1}\phi_j' \in C\left(R \times \bigcup_{\mu\geq 0} \{(\mu,\zeta)\ |\ \zeta \in L(c(-\infty)\mu)\}\right)$$

for j = 3, 4. Moreover

$$\phi_1,\ \rho^{-1}\phi_1' \in C\left\{R \times \bigcup_{\mu\geq 0} \{(\mu,\zeta) \mid \zeta \in R^+(c(\infty)\mu)\}\right\},$$

(3.14)

$$\phi_2,\ \rho^{-1}\phi_2' \in C\left\{R \times \bigcup_{\mu\geq 0} \{(\mu,\zeta) \mid \zeta \in R^-(c(\infty)\mu)\}\right\},$$

$$\phi_3,\ \rho^{-1}\phi_3' \in C\left\{R \times \bigcup_{\mu\geq 0} \{(\mu,\zeta) \mid \zeta \in R^-(c(-\infty)\mu)\}\right\},$$

$$\phi_4,\ \rho^{-1}\phi_4' \in C\left\{R \times \bigcup_{\mu\geq 0} \{(\mu,\zeta) \mid \zeta \in R^+(c(-\infty)\mu)\}\right\}.$$

Corollary 3.3. For each fixed $(y,\mu) \in R \times R_+$ the mappings

(3.15) $$\zeta \to \phi_j(y,\mu,\zeta),\ \zeta \to \rho^{-1}(y)\ \phi_j'(y,\mu,\zeta)$$

are analytic for

$$j = 1,\ \zeta \in L(c(\infty)\mu) \cup R^+(c(\infty)\mu)^{\text{int}},$$

(3.16)

$$j = 2,\ \zeta \in L(c(\infty)\mu) \cup R^-(c(\infty)\mu)^{\text{int}},$$

$$j = 3,\ \zeta \in L(c(-\infty)\mu) \cup R^-(c(-\infty)\mu)^{\text{int}},$$

$$j = 4,\ \zeta \in L(c(-\infty)\mu) \cup R^+(c(-\infty)\mu)^{\text{int}},$$

where $R^{\pm}(\kappa)^{\text{int}} = R(\kappa) \cap \{\zeta \mid \pm\text{Im}\ \zeta > 0\}$.

Corollary 3.4. The asymptotic estimates for ϕ_j and $\rho^{-1}\phi_j'$ of Theorem 3.1 hold uniformly for (μ,ζ) in any compact set Γ_j such that for

$$j = 1,\ \Gamma_1 \subset \bigcup_{\mu\geq 0} \{(\mu,\zeta) \mid \zeta \in L(c(\infty)\mu) \cup R^+(c(\infty)\mu)\},$$

(3.17)

$$j = 2,\ \Gamma_2 \subset \bigcup_{\mu\geq 0} \{(\mu,\zeta) \mid \zeta \in L(c(\infty)\mu) \cup R^-(c(\infty)\mu)\},$$

$$j = 3,\ \Gamma_3 \subset \bigcup_{\mu\geq 0} \{(\mu,\zeta) \mid \zeta \in L(c(-\infty)\mu) \cup R^-(c(-\infty)\mu)\},$$

$$j = 4,\ \Gamma_4 \subset \bigcup_{\mu\geq 0} \{(\mu,\zeta) \mid \zeta \in L(c(-\infty)\mu) \cup R^+(c(-\infty)\mu)\}.$$

The special solutions $\phi_j(y,\mu,\zeta)$ are not, in general, uniquely determined by the asymptotic conditions (3.7), (3.8), (3.10), (3.11). Indeed, if Re $q_+'(\mu,\lambda) > 0$ (resp., Re $q_+'(\mu,\lambda) < 0$) it is clear that any multiple of

ϕ_2 (resp., ϕ_1) can be added to ϕ_1 (resp., ϕ_2). A similar remark holds for ϕ_3 and ϕ_4. However, for each $\zeta \in C$ a sub-dominant solution (one with minimal growth at $y = \infty$ or $y = -\infty$) is unique. In particular, since

$$\text{(3.18)} \qquad \begin{aligned} &\text{Re } q_\pm'(\mu,\zeta) > 0 \text{ for } \zeta \in L(c(\pm\infty)\mu) \\ &\text{Re } q_\pm'(\mu,\zeta) \geq 0 \text{ for } \zeta \in R^-(c(\pm\infty)\mu) \\ &\text{Re } q_\pm'(\mu,\zeta) \leq 0 \text{ for } \zeta \in R^+(c(\pm\infty)\mu) \end{aligned}$$

one can prove

Corollary 3.5. The solution ϕ_2 is uniquely determined by (3.8) for all $\zeta \in L(c(\infty)\mu) \cup R^-(c(\infty)\mu)$. Similarly, ϕ_3 is uniquely determined by (3.10) for $\zeta \in L(c(-\infty)\mu) \cup R^-(c(-\infty)\mu)$, ϕ_1 is uniquely determined by (3.7) in $R^+(c(\infty)\mu)$ and ϕ_4 is uniquely determined by (3.11) in $R^+(c(-\infty)\mu)$.

When $\text{Re } \zeta = c^2(\pm\infty)\mu^2$ Theorem 3.1 provides no information about the asymptotic behavior for $y \to \pm\infty$ of solutions of $A_\mu\phi = \zeta\phi$. However, positive results can be obtained by strengthening hypothesis (1.1). The following extension of a known result [14, p. 209] will be used in §4.

Theorem 3.6. Assume that $\rho(y)$ and $c(y)$ satisfy hypothesis (2.1.4) and

$$\text{(3.19)} \qquad \int_0^\infty |\rho(y) - \rho(\infty)| \, dy < \infty, \quad \int_0^\infty y^2 \, |c(y) - c(\infty)| \, dy < \infty.$$

Then there exist functions

$$\text{(3.20)} \qquad \phi_j : R \times R_+ \to R, \; j = 1,2,$$

such that for every $\mu \in R_+$ the pair $\phi_1(y,\mu)$, $\phi_2(y,\mu)$ is a solution basis for $A_\mu\phi = c^2(\infty)\mu^2\phi$,

$$\text{(3.21)} \qquad \left.\begin{aligned} \phi_1(y,\mu) &= 1 + o(1) \\ \rho^{-1}(y)\,\phi_1'(y,\mu) &= o(1) \end{aligned}\right\} \; y \to +\infty,$$

and

$$\text{(3.22)} \qquad \left.\begin{aligned} \phi_2(y,\mu) &= \rho(\infty)y[1 + o(1)] \\ \rho^{-1}(y)\,\phi_2'(y,\mu) &= 1 + o(1) \end{aligned}\right\} \; y \to +\infty.$$

Lagrange's formula for A_μ may be written

$$\int_{y_1}^{y_2} \{\psi A_\mu \phi - \phi A_\mu \psi\}\, c^{-2}(y)\rho^{-1}(y)\,dy = [\phi\psi](y_2) - [\phi\psi](y_1) \tag{3.23}$$

where

$$[\phi\psi](y) = \phi(y)(\rho^{-1}(y)\ \psi'(y)) - \psi(y)(\rho^{-1}(y)\ \phi'(y)). \tag{3.24}$$

In particular, if ϕ and ψ are solutions of $A_\mu\phi = \zeta\phi$, $A_\mu\psi = \zeta\psi$ on an interval I then $[\phi\psi](y) = \text{const.}$ on I and $[\phi\psi](y) = 0$ on I if and only if ϕ and ψ are linearly dependent there. By combining these facts and Theorem 3.1 one can show that

$$\begin{aligned} [\phi_1(\cdot,\mu,\zeta)\ \phi_2(\cdot,\mu,\zeta)] &= -2\ \rho^{-1}(\infty)\ q_+'(\mu,\zeta), \\ [\phi_3(\cdot,\mu,\zeta)\ \phi_4(\cdot,\mu,\zeta)] &= -2\ \rho^{-1}(-\infty) q_-'(\mu,\zeta), \end{aligned} \tag{3.25}$$

which imply

Corollary 3.7. The pair $\phi_1(y,\mu,\zeta)$, $\phi_2(y,\mu,\zeta)$ is a solution basis for $A_\mu\phi = \zeta\phi$ for all $(\mu,\zeta) \in R_+ \times (L(c(\infty)\mu) \cup R(c(\infty)\mu))$. Similarly, the pair $\phi_3(y,\mu,\zeta)$, $\phi_4(y,\mu,\zeta)$ is a solution basis for all $(\mu,\zeta) \in R_+ \times (L(c(-\infty)\mu) \cup R(c(-\infty)\mu))$.

This completes the formulation of the results of §3 and the proofs will now be given. The method of proof involves replacing $A_\mu\phi = \zeta\phi$ by an equivalent first order system. The latter can be regarded as a perturbation of the corresponding limit systems for $y \to \pm\infty$. In this way integral equations are established for solutions with prescribed asymptotic behavior for $y \to \infty$ or $y \to -\infty$ and these equations are solved by classical Banach space methods. This technique for constructing solutions with prescribed asymptotic behavior is well known - see for example [5, p. 1408] and [16, Ch. VII].

A first order system equivalent to $A_\mu\phi = \zeta\phi$. If $\phi(y)$ is any solution of (3.1) on an interval I and if

$$\begin{cases} \psi_1(y) = \phi(y) \\ \psi_2(y) = \rho^{-1}(y)\ \phi'(y) \end{cases} \tag{3.26}$$

then $\psi_1, \psi_2 \in AC(I)$ (cf. (3.2)) and

$$(3.27)\qquad \begin{cases} \psi_1'(y) = \rho(y)\, \psi_2(y) \\ \psi_2'(y) = \rho^{-1}(y)[\mu^2 - \zeta\, c^{-2}(y)]\, \psi_1(y) \end{cases}$$

for almost every $y \in I$. Thus the column vector $\psi(y)$ with components $\psi_1(y)$, $\psi_2(y)$ is a solution of the first order linear system

$$(3.28)\qquad \psi'(y) = M(y,\mu,\zeta)\, \psi(y)$$

where

$$(3.29)\qquad M(y,\mu,\zeta) = \begin{bmatrix} 0 & \rho(y) \\ \rho^{-1}(y)[\mu^2 - \zeta\, c^{-2}(y)] & 0 \end{bmatrix}.$$

Conversely, if $\psi \in AC(I)$ is a solution of (3.28), (3.29) and if $\phi(y) = \psi_1(y)$ then ϕ is a solution of (3.1). The solutions of Theorem 3.1 will be constructed by integrating (3.28), (3.29).

<u>The limit system for $y \to +\infty$ and its solutions</u>. By replacing $\rho(y)$, $c(y)$ in (3.28), (3.29) by $\rho(\infty)$, $c(\infty)$ one obtains the system

$$(3.30)\qquad \psi'(y) = M_0(\mu,\zeta)\, \psi(y)$$

where

$$(3.31)\qquad M_0(\mu,\zeta) = \begin{bmatrix} 0 & \rho(\infty) \\ \rho^{-1}(\infty)[\mu^2 - \zeta\, c^{-2}(\infty)] & 0 \end{bmatrix}.$$

$M_0(\mu,\zeta)$ has distinct eigenvalues $q_+'(\mu,\zeta)$, $-q_+'(\mu,\zeta)$ for $\zeta \in L(c(\infty)\mu) \cup R(c(\infty)\mu)$. The columns of

$$(3.32)\qquad B(\mu,\zeta) = \begin{bmatrix} 1 & 1 \\ \rho^{-1}(\infty)\, q_+'(\mu,\zeta) & -\rho^{-1}(\infty)\, q_+'(\mu,\zeta) \end{bmatrix}$$

are corresponding eigenvectors. Hence

$$(3.33)\qquad M_0(\mu,\zeta)\, B(\mu,\zeta) = B(\mu,\zeta)\, D(\mu,\zeta)$$

where

$$D(\mu,\zeta) = \begin{pmatrix} q_+'(\mu,\zeta) & 0 \\ 0 & -q_+'(\mu,\zeta) \end{pmatrix}. \tag{3.34}$$

System (3.30), (3.31) may be integrated by the substitution

$$\psi = B(\mu,\zeta)z. \tag{3.35}$$

It follows that $z'(y) = D(\mu,\zeta)\, z(y)$, whence

$$\begin{cases} z_1(y) = c_1 \exp\{q_+'(\mu,\zeta)y\} \\ z_2(y) = c_2 \exp\{-q_+'(\mu,\zeta)y\} \end{cases} \tag{3.36}$$

and therefore

$$\begin{cases} \psi_1(y) = c_1 \exp\{q_+'(\mu,\zeta)y\} + c_2 \exp\{-q_+'(\mu,\zeta)y\} \\ \psi_2(y) = \rho^{-1}(\infty)\, q_+'(\mu,\zeta)(c_1 \exp\{q_+'(\mu,\zeta)y\} - c_2 \exp\{-q_+'(\mu,\zeta)y\}) \end{cases} \tag{3.37}$$

where c_1, c_2 are constants of integration.

Application of perturbation theory. System (3.28) may be regarded as a perturbation of the limit system (3.30). Thus if $N(y,\mu,\zeta)$ is defined by

$$M(y,\mu,\zeta) = M_0(\mu,\zeta) + N(y,\mu,\zeta) \tag{3.38}$$

then

$$N(y,\mu,\zeta) = \begin{pmatrix} 0 & a_1(y) \\ \mu^2\, a_2(y) + \zeta\, a_3(y) & 0 \end{pmatrix} \tag{3.39}$$

where

$$\begin{cases} a_1(y) = \rho(y) - \rho(\infty) \\ a_2(y) = \rho^{-1}(y) - \rho^{-1}(\infty) \\ a_3(y) = -[\rho^{-1}(y)\, c^{-2}(y) - \rho^{-1}(\infty)\, c^{-2}(\infty)]. \end{cases} \tag{3.40}$$

Note that each of these functions is in $L_1(y_0,\infty)$ for every $y_0 \in R$. For $a_1(y)$ this is part of hypothesis (1.1). For $a_2(y)$ and $a_3(y)$ it follows from (2.1.4) and (1.1). For example, one can write

$$a_2(y) = (-\rho^{-1}(\infty)\ \rho^{-1}(y))(\rho(y) - \rho(\infty)) \tag{3.41}$$

which exhibits a_2 as a product of a bounded measurable function and a function in $L_1(y_0,\infty)$.

On combining (3.28), (3.38) and making the substitution (3.35), one finds that (3.28) is equivalent to the system

$$z'(y) = D(\mu,\zeta)\ z(y) + E(y,\mu,\zeta)\ z(y) \tag{3.42}$$

where

$$E(y,\mu,\zeta) = B^{-1}(\mu,\zeta)\ N(y,\mu,\zeta)\ B(\mu,\zeta) \tag{3.43}$$

has components that are in $L_1(y_0,\infty)$. Solutions of (3.42) will be constructed which are asymptotically equal, for $y \to +\infty$, to the solutions (3.36) of $z' = D(\mu,\zeta)z$.

Proof of Theorem 3.1. The proof will be given for the function ϕ_1 only. The remaining cases can be proved by the same method. Solutions of (3.42) are related to the corresponding solutions of (3.1) by

$$\begin{cases} \phi = z_1 + z_2 \\ \rho^{-1}\phi' = \rho^{-1}(\infty)\ q_+'(z_1 - z_2),\ q_+' = q_+'(\mu,\zeta). \end{cases} \tag{3.44}$$

Thus ϕ will be a solution of (3.1) that satisfies (3.7) if z is a solution of (3.42) that satisfies

$$z_1 = \exp\{q_+'y\}\eta_1,\ z_2 = \exp\{q_+'y\}\eta_2 \tag{3.45}$$

and

$$\eta_1(y) = 1 + o(1),\ \eta_2(y) = o(1) \text{ for } y \to \infty. \tag{3.46}$$

Equations (3.42) and (3.45) imply

$$(3.47)\quad \begin{cases} \eta_1' = & E_{11}\,\eta_1 + E_{12}\,\eta_2, \\ \eta_2' = -2\,q_+'\,\eta_2 & + E_{21}\,\eta_1 + E_{22}\,\eta_2, \end{cases}$$

and hence by integration

$$(3.48)\quad \begin{aligned} \eta_1(y) &= c_1 + \int_{y_0}^{y} E_{1j}(y')\,\eta_j(y')dy' \\ \eta_2(y) &= \exp\{-2\,q_+'y\}\,c_2 + \int_{y_1}^{y} \exp\{-2\,q_+'(y-y')\}\,E_{2j}(y')\,\eta_j(y')dy' \end{aligned}$$

where c_1, c_2, y_0, y_1 are constants and the summation convention has been used (j is summed over $j = 1,2$).

Construction of ϕ_1 for $\zeta \in L(c(\infty)\mu)$. By (3.18), Re $q_+'(\mu,\zeta) > 0$ for all $\zeta \in L(c(\infty)\mu)$. Thus to construct a solution of (3.47) that satisfies (3.46) it is natural to choose $c_1 = 1$, $c_2 = 0$, $y_0 = +\infty$ and y_1 finite in (3.48). This gives the system of integral equations

$$(3.49)\quad \left.\begin{aligned} \eta_1(y) &= 1 - \int_{y}^{\infty} E_{1j}(y')\,\eta_j(y')dy' \\ \eta_2(y) &= \int_{y_1}^{y} \exp\{-2\,q_+'(y-y')\}\,E_{2j}(y')\,\eta_j(y')dy' \end{aligned}\right\}\; y \ge y_1.$$

It is natural to study system (3.49) in the space

$$(3.50)\quad X = CB([y_1,\infty),C^2)$$

of two-component vector functions of y whose components are continuous and bounded on $y_1 \le y < \infty$. X is a Banach space with norm

$$(3.51)\quad \|\eta\| = \sup_{y \ge y_1} (|\eta_1(y)| + |\eta_2(y)|).$$

The system has the form

$$(3.52)\quad \eta(y) = \eta^0 + \int_{y_1}^{\infty} K(y,y')\,\eta(y')dy',\; y \ge y_1,$$

where $\eta(y)$ and η^0 are column vectors with components $(\eta_1(y),\eta_2(y))$ and $(1,0)$, respectively, and the matrix kernel $K(y,y')$ is defined by

$$(3.53)\qquad K_{1j}(y,y') = \begin{cases} 0, & y_1 \le y' < y, \\ -E_{1j}(y'), & y_1 \le y \le y' \end{cases}$$

$$(3.54)\qquad K_{2j}(y,y') = \begin{cases} \exp\{-2\, q_+'(y-y')\}\, E_{2j}(y'), & y_1 \le y' < y, \\ 0, & y_1 \le y \le y', \end{cases}$$

and $j = 1,2$. The conditions $E_{jk} \in L_1(y_1,\infty)$ and $\mathrm{Re}\, q_+' > 0$ imply that the operator K defined by (3.52), (3.53) and (3.54) maps X into itself. To show that K is a bounded operator in X and estimate its norm note that

$$(3.55)\qquad |(K\eta)_j(y)| \le \|\eta\| \int_{y_1}^{\infty} (|E_{j1}(y)| + |E_{j2}(y)|)\, dy$$

for $j = 1,2$ and all $y \ge y_1$. It follows that

$$(3.56)\qquad \|K\| \le \int_{y_1}^{\infty} \sum_{j,k=1}^{2} |E_{jk}(y)|\, dy.$$

In particular, since $E_{jk} \in L_1(y_0,\infty)$ for every $y_0 \in R$, (3.56) implies that $\|K\| < 1$ for every sufficiently large y_1. For such a value of y_1 the equation

$$(3.57)\qquad \eta = \eta^0 + K\eta$$

has a unique solution $\eta \in X$ given by

$$(3.58)\qquad \eta = \sum_{m=0}^{\infty} K^m \eta^0.$$

Moreover, (3.57), or equivalently (3.49), implies that $\eta_1(y)$, $\eta_2(y)$ satisfy (3.47) for $y \ge y_1$. These functions then have unique continuations to solutions of (3.47) for all $y \in R$, by the classical existence and uniqueness theory for linear systems.

Of course, η_1 and η_2 are functions of μ and ζ as well as y because q_+' and the E_{jk} depend on these variables. The solution ϕ_1 of Theorem 3.1 will be defined by

$$(3.59)\qquad \phi_1(y,\mu,\zeta) = \exp\{q_+'(\mu,\zeta)y\}\,(\eta_1(y,\mu,\zeta) + \eta_2(y,\mu,\zeta)).$$

To complete the proof that ϕ_1 is the desired function on $R \times R_+ \times L(c(\infty)\mu)$

it is only necessary to verify that (3.46) holds for each $(\mu,\zeta) \in R_+ \times L(c(\infty)\mu)$. It is clear from (3.49) that

$$|\eta_1(y) - 1| \le \|\eta\| \int_{y_1}^{\infty} \sum_{j=1}^{2} |E_{1j}(y')| \, dy' = o(1), \; y \to +\infty. \tag{3.60}$$

For η_2, (3.49) implies that

$$|\eta_2(y)| \le \|\eta\| \left\{ \int_{y_1}^{y_2} \exp\{-2q_+'(y-y')\} \sum_{j=1}^{2} |E_{2j}(y')| dy' + \int_{y_2}^{\infty} \sum_{j=1}^{2} |E_{2j}(y')| dy' \right\} \tag{3.61}$$

for every $y_2 \ge y_1$ and every $y \ge y_2$. Hence for any fixed y_2 one has

$$\limsup_{y\to\infty} |\eta_2(y)| \le \|\eta\| \int_{y_2}^{\infty} \sum_{j=1}^{2} |E_{2j}(y')| \, dy' \tag{3.62}$$

because $\mathrm{Re}\, q_+' > 0$. Since y_2 in (3.62) is arbitrary it follows that $\eta_2(y) = o(1)$.

Construction of ϕ_1 for $\zeta \in R(c(\infty)\mu)$. By (3.3) $R(c(\infty)\mu)$ has the decomposition

$$R(c(\infty)\mu) = R^+(c(\infty)\mu) \cup R^-(c(\infty)\mu)^{\mathrm{int}}. \tag{3.63}$$

Moreover, for $\zeta \in R^-(c(\infty)\mu)^{\mathrm{int}}$ one has $\mathrm{Re}\, q_+'(\mu,\zeta) > 0$ and hence the construction of the preceding case is valid. In the complementary case where $\zeta \in R^+(c(\infty)\mu)$ one has $\mathrm{Re}\, q_+'(\mu,\zeta) \le 0$ and it is permissible to take $c_1 = 1$, $c_2 = 0$, $y_0 = y_1 = \infty$ in (3.48). The resulting system of integral equations

$$\left.\begin{aligned} \eta_1(y) &= 1 - \int_y^{\infty} E_{1j}(y')\, \eta_j(y')dy' \\ \eta_2(y) &= -\int_y^{\infty} \exp\{-2q_+'(y-y')\}\, E_{2j}(y')\, \eta_j(y')dy' \end{aligned}\right\} \; y \ge y_1 \tag{3.64}$$

again defines an equation (3.57) in the Banach space X. Moreover, $|\exp\{-2q_+'(y-y')\}| \le 1$ for $y \le y' < \infty$ and (3.56) is again valid. It follows that for y_1 large enough (3.64) has a unique solution given by (3.58). The solution has a unique continuation to a solution of (3.47) on the interval $y \in R$. The validity of the asymptotic condition (3.46) is obvious from (3.64); cf. (3.60).

Proof of Corollary 3.2. Again the proof will be given for ϕ_1 only. Note that by (3.39)

$$N(y,\mu,\zeta) = N_1(y) + \mu^2 N_2(y) + \zeta N_3(y) \tag{3.65}$$

where the components of $N_j(y)$ are in $L_1(y_0,\infty)$ for $j = 1,2,3$ and every $y_0 \in R$. Thus by (3.43)

$$\begin{aligned} E(y,\mu,\zeta) &= B^{-1}(\mu,\zeta)\, N_1(y)\, B(\mu,\zeta) + \mu^2\, B^{-1}(\mu,\zeta)\, N_2(y)\, B(\mu,\zeta) \\ &\quad + \zeta\, B^{-1}(\mu,\zeta)\, N_3(y)\, B(\mu,\zeta). \end{aligned} \tag{3.66}$$

Proof of (3.12) for ϕ_1. Note that $q_+^!(\mu,\zeta)$, and hence also $B(\mu,\zeta)$ and $B^{-1}(\mu,\zeta)$ are continuous functions on the set

$$\bigcup_{\mu>0} \{(\mu,\zeta) \mid \zeta \in L(c(\infty)\mu)\}. \tag{3.67}$$

Thus by using the estimate (3.56) for the operator $K = K(\mu,\zeta)$ in X one can show that for each compact subset Γ of the set (3.67) and each $\delta < 1$ there is a constant $y_1 = y_1(\Gamma,\delta)$ such that, taking $y_1 = y_1(\Gamma,\delta)$ in the definition of $K(\mu,\zeta)$, one has

$$\|K(\mu,\zeta)\| \le \delta \text{ for all } (\mu,\zeta) \in \Gamma. \tag{3.68}$$

Hence the series (3.58) converges uniformly in X for $(\mu,\zeta) \in \Gamma$ which implies the continuity of ϕ_1 and $\rho^{-1}\phi_1'$ on the set $[y_1,\infty) \times \Gamma$. Their continuity on $R \times \Gamma$ then follows from the classical theorem on the continuous dependence of solutions of initial value problems on parameters. This implies the result (3.12) for ϕ_1 because Γ was an arbitrary compact subset of the set (3.67).

Proof of (3.14) for ϕ_1. The method used in the preceding case is applicable to the operator $K(\mu,\zeta)$ in X defined by (3.64).

Remark on Corollary 3.2. The argument given above can also be used to show that

$$\phi_1,\ \rho^{-1}\phi_1' \in C\left(R \times \bigcup_{\mu>0} \{(\mu,\zeta) \mid \zeta \in R^-(c(\infty)\mu)^{\text{int}}\}\right). \tag{3.69}$$

However, the continuity of ϕ_1 and $\rho^{-1}\phi_1'$ on the set

$$R \times \bigcup_{\mu>0} \{(\mu,\zeta) \mid \zeta \in R(c(\infty)\mu)\} \tag{3.70}$$

cannot be asserted since the constructions for $\zeta \in R^+(c(\infty)\mu)$ and

$\zeta \in R^-(c(\infty)\mu)^{\text{int}}$ are different. Indeed, continuity of ϕ_1 on the set (3.70) is not to be expected since ϕ_1 is not uniquely determined when $\zeta \in R^-(c(\infty)\mu)^{\text{int}}$.

Proof of Corollary 3.3. The components of the matrix-valued function $E(y,\mu,\zeta)$ are analytic functions of $\zeta \in L(c(\infty)\mu) \cup R^+(c(\infty)\mu)^{\text{int}}$ for fixed values of y, μ. Hence the uniform convergence of the Neumann series (3.58) on compact subsets of this set, which follows from the proof of Corollary 3.2, implies the validity of Corollary 3.3 for ϕ_1. The remaining cases can be proved by the same method.

Proof of Corollary 3.4. The proof will be given for the case of ϕ_1 and (μ,ζ) in a compact subset Γ of the set (3.67). The remaining cases can be proved similarly. $\phi_1(y,\mu,\zeta)$ was defined by (3.59) and the functions $\eta_j(y,\mu,\zeta)$ satisfy

$$\begin{aligned} \eta_1(y,\mu,\zeta) - 1 &= -\int_y^\infty E_{1j}(y',\mu,\zeta)\, \eta_j(y',\mu,\zeta)dy', \\ \eta_2(y,\mu,\zeta) &= -\int_y^\infty \exp\{-2q_+'(\mu,\zeta)(y-y')\}\, E_{2j}(y',\mu,\zeta)\, \eta_j(y',\mu,\zeta)dy'. \end{aligned} \tag{3.71}$$

It must be shown that these integrals tend to zero when $y \to \infty$, uniformly for $(\mu,\zeta) \in \Gamma$. Now (3.58) and the estimate (3.68) from the proof of Corollary 3.2 imply that $\|\eta(\cdot,\mu,\zeta)\| \le (1-\delta)^{-1}$ for all $(\mu,\zeta) \in \Gamma$. It follows that for fixed $y' \ge y \ge y_1(\Gamma,\delta)$ one has

$$\begin{aligned} |E_{kj}(y',\mu,\zeta)\, \eta_j(y',\mu,\zeta)| &\le \|E(y',\mu,\zeta)\|\, \|\eta(y',\mu,\zeta)\| \\ &\le (1-\delta)^{-1}\, \|E(y',\mu,\zeta)\| \end{aligned} \tag{3.72}$$

for $(\mu,\zeta) \in \Gamma$. Now the continuity of $B(\mu,\zeta)$ implies that there is a $\gamma = \gamma(\Gamma)$ such that

$$\|B(\mu,\zeta)\|\, \|B^{-1}(\mu,\delta)\|\, (1 + \mu^2 + |\zeta|) \le \gamma \text{ for } (\mu,\zeta) \in \Gamma. \tag{3.73}$$

It follows from (3.66) that

$$\|E(y',\mu,\zeta)\| \le \gamma \sum_{j=1}^{3} \|N_j(y')\|, \quad (\mu,\zeta) \in \Gamma. \tag{3.74}$$

Combining (3.71), (3.72) and (3.74) gives

$$|\eta_1(y,\mu,\zeta) - 1| \le \gamma(1-\delta)^{-1} \int_y^\infty \sum_{j=1}^{3} \|N_j(y')\|\, dy' \tag{3.75}$$

for all $y \geq y_1(\Gamma,\delta)$ and $(\mu,\zeta) \in \Gamma$. Since each $N_j \in L_2(y_0,\infty)$, (3.75) implies that $\eta_1(y,\mu,\zeta) - 1 = o(1)$ uniformly for $(\mu,\zeta) \in \Gamma$.

The case of $\eta_2(y,\mu,\zeta)$ is more complicated. Note that

$$(3.76)\quad |\eta_2(y,\mu,\zeta)| \leq \gamma(1-\delta)^{-1} \int_y^\infty \exp\{-2\,\mathrm{Re}\, q_+'(\mu,\zeta)(y-y')\} \sum_{j=1}^3 \|N_j(y')\|\, dy'.$$

Now the continuity of $q_+'(\mu,\zeta)$ and the definition of $L(c(\infty)\mu)$ imply that there is a $\kappa = \kappa(\Gamma) > 0$ such that

$$(3.77)\qquad 2\,\mathrm{Re}\, q_+'(\mu,\zeta) \geq \kappa > 0 \text{ for all } (\mu,\zeta) \in \Gamma.$$

Combining (3.76), (3.77) one has, if $\gamma_1 = \gamma(1-\delta)^{-1}$,

$$(3.78)\quad \begin{aligned} |\eta_2(y,\mu,\zeta)| &\leq \gamma_1 \left\{\int_y^{y_2} \exp\{-\kappa(y-y')\} \sum_{j=1}^3 \|N_j(y')\|dy' + \int_{y_2}^\infty \sum_{j=1}^3 \|N_j(y')\|dy'\right\} \\ &\leq \gamma_1 \left\{\exp\{-\kappa(y-y_2)\} \int_{y_1}^\infty \sum_{j=1}^3 \|N_j(y')\|dy' + \int_{y_2}^\infty \sum_{j=1}^3 \|N_j(y')\|dy'\right\} \end{aligned}$$

for all $y_2 \geq y_1(\Gamma,\delta)$, $y \geq y_2$ and $(\mu,\zeta) \in \Gamma$.

Now let $\varepsilon > 0$ be given and choose $y_2 = y_2(\varepsilon,\Gamma,\delta) \geq y_1(\Gamma,\delta)$ such that

$$(3.79)\qquad \gamma_1 \int_{y_2}^\infty \sum_{j=1}^3 \|N_j(y')\|dy' < \varepsilon/2,$$

and hence

$$(3.80)\quad |\eta_2(y,\mu,\zeta)| \leq \gamma_1 \exp\{-\kappa(y-y_2)\} \int_{y_1}^\infty \sum_{j=1}^3 \|N_j(y')\|dy' + \varepsilon/2$$

for all $y \geq y_2(\varepsilon,\Gamma,\delta) \geq y_1(\Gamma,\delta)$ and $(\mu,\zeta) \in \Gamma$. Finally, choose a $y_3(\varepsilon,\Gamma,\delta) \geq y_2(\varepsilon,\Gamma,\delta)$ such that

$$(3.81)\qquad \gamma_1 \exp\{-\kappa(y-y_2)\} \int_{y_1}^\infty \sum_{j=1}^3 \|N_j(y')\|dy' < \varepsilon/2.$$

for all $y \geq y_3(\varepsilon,\Gamma,\delta)$. It follows from (3.80) and (3.81) that $|\eta_2(y,\mu,\zeta)| < \varepsilon$ for all $y \geq y_3(\varepsilon,\Gamma,\delta)$ and $(\mu,\zeta) \in \Gamma$; i.e., $\eta_2(y,\mu,\zeta) = o(1)$ uniformly for $(\mu,\zeta) \in \Gamma$.

Proof of Corollary 3.5. It will be shown that $\phi_1(y,\mu,\zeta)$ is uniquely determined by (3.1) and (3.7) when $\zeta \in R^+(c(\infty)\mu)$. The other cases are proved similarly.

Assume that for some $\zeta \in R^+(c(\infty)\mu)$ there are two solutions of (3.1), (3.7). Then their difference $\phi(y)$ would satisfy (3.1) and $\phi(y) = o(1)$, $\rho^{-1}(y)\,\phi'(y) = o(1)$ because $\operatorname{Re} q_+' \le 0$ for $\zeta \in R^+(c(\infty)\mu)$. It follows that the corresponding pair $\eta_1(y)$, $\eta_2(y)$, defined by (3.44) and (3.45), would necessarily satisfy

$$\begin{aligned} \eta_1(y) &= -\int_y^\infty E_{1j}(y')\,\eta_j(y')dy' \\ \eta_2(y) &= -\int_y^\infty \exp\{-2q_+'(y-y')\}\,E_{2j}(y')\,\eta_j(y')dy' \end{aligned} \tag{3.82}$$

since $|\exp\{-2q_+'(y-y')| \le 1$ for $y \le y'$. But (3.82) is equivalent to the equation $\eta = K\eta$ in X. If y_1 is chosen so large that $\|K\| < 1$ then $\eta = K\eta$ has the unique solution $\eta(y) \equiv 0$ for $y \ge y_1$. The unique continuation of this solution of (3.47) is then zero for all $y \in R$. Thus $\phi(y) \equiv 0$ for $y \in R$, which proves the uniqueness.

Proof of Theorem 3.6. The equation $A_\mu\phi = c^2(\infty)\mu^2\phi$ is equivalent under the mapping (3.26) with the system (see (3.31), (3.38))

$$\begin{cases} \psi_1' = \rho(\infty)\psi_2 + B_1(y)\psi_2 \\ \psi_2' = B_2(y)\psi_1 \end{cases} \tag{3.83}$$

where $B_1(y) = a_1(y) = \rho(y) - \rho(\infty)$ and

$$\begin{aligned} B_2(y) &= \mu^2\,\rho^{-1}(y)(1 - c^2(\infty)c^{-2}(y)) \\ &= \mu^2\,\rho^{-1}(y)\,c^{-2}(y)[c(y) + c(\infty)][c(y) - c(\infty)]. \end{aligned} \tag{3.84}$$

It follows from hypotheses (2.1.4) and (3.19) that

$$B_1(y),\ B_2(y),\ y\,B_2(y),\ y^2\,B_2(y) \in L_1(y_0,\infty) \tag{3.85}$$

for every $y_0 \in R$ and every $\mu \ge 0$.

Construction of ϕ_1. Application of the variation of constants formula to the system (3.83) gives the integrated form

$$\begin{aligned} \psi_1(y) &= c_1 + \rho(\infty)\,c_2 y + \int_{y_0}^{y} \{\rho(\infty)(y-y')\,B_2(y')\,\psi_1(y') + B_1(y')\,\psi_2(y')\}dy' \\ \psi_2(y) &= c_2 + \int_{y_0}^{y} B_2(y')\,\psi_1(y')dy'. \end{aligned} \tag{3.86}$$

Now ϕ_1 will satisfy $A_\mu\phi_1 = c^2(\infty)\mu^2\phi_1$ and the asymptotic condition (3.21) provided that $\psi_1 = \phi_1$, $\psi_2 = \rho^{-1}\phi_1'$ satisfies (3.86) and

$$\psi_1(y) = 1 + o(1),\ \psi_2(y) = o(1),\ y \to \infty. \tag{3.87}$$

To construct such a solution take $c_1 = 1$, $c_2 = 0$ and $y_0 = \infty$ in (3.86). This gives the system

$$\begin{cases} \psi_1(y) = 1 - \displaystyle\int_y^\infty \{\rho(\infty)(y-y')\, B_2(y')\, \psi_1(y') + B_1(y')\, \psi_2(y')\}\, dy' \\ \psi_2(y) = -\displaystyle\int_y^\infty B_2(y')\, \psi_1(y')\, dy' \end{cases} \tag{3.88}$$

or

$$\psi(y) = \psi^0 + \int_{y_1}^\infty K(y,y')\,\psi(y')\, dy', \quad y \geq y_1, \tag{3.89}$$

where $\psi(y)$ and ψ^0 have components $\psi_1(y)$, $\psi_2(y)$ and 1, 0 respectively, $K(y,y') \equiv 0$ for $y \geq y'$ and

$$\begin{aligned} K_{11}(y,y') &= \rho(\infty)(y'-y)\, B_2(y') \\ K_{12}(y,y') &= -B_1(y') \\ K_{21}(y,y') &= -B_2(y') \\ K_{22}(y,y') &= 0 \end{aligned} \tag{3.90}$$

for $y \leq y'$. As in the proof of Theorem 3.1, one has

$$|(K\psi)_j(y)| \leq \|\psi\| \int_{y_1}^\infty (|K_{j1}(y,y')| + |K_{j2}(y,y')|)\, dy' \tag{3.91}$$

and hence

$$\begin{aligned} |(K\psi)_1(y)| &\leq \|\psi\| \left\{\rho(\infty) \int_y^\infty (y+y')\ |B_2(y')|dy' + \int_y^\infty |B_1(y')|dy'\right\} \\ &\leq \|\psi\| \left\{2\rho(\infty) \int_y^\infty y'\ |B_2(y')|dy' + \int_y^\infty |B_1(y')|dy'\right\} \end{aligned} \tag{3.92}$$

and

$$(3.93)\qquad |(K\psi)_2(y)| \le \|\psi\| \int_y^\infty |B_2(y')|\, dy' .$$

In particular,

$$(3.94)\qquad \|K\| \le 2\rho(\infty) \int_{y_1}^\infty y' \,|B_2(y')|dy' + \int_{y_1}^\infty |B_1(y')|dy' + \int_{y_1}^\infty |B_2(y')|dy'.$$

Thus (3.85) implies that K is contractive in X for y_1 large enough. Hence (3.88) has a unique solution on $[y_1,\infty)$ which can be continued as a solution of (3.83) to all $y \in R$. Moreover, (3.88), (3.92) and (3.93) imply that $|\psi_1(y) - 1| = |(K\psi)_1(y)| = o(1)$ and $|\psi_2(y)| = |(K\psi)_2(y)| = o(1)$. In fact (3.85) implies that $|\psi_2(y)| = o(y^{-2})$. Thus (3.87) is satisfied.

<u>Construction of ϕ_2</u>. ϕ_2 will satisfy $A_\mu\phi_2 = c^2(\infty)\mu^2\phi_2$ and the asymptotic condition (3.22) provided

$$(3.95)\qquad \begin{cases} \psi_1(y) = \phi_2(y) = \rho(\infty)y\,\eta_1(y) \\ \psi_2(y) = \rho^{-1}(y)\phi_2'(y) = \eta_2(y) \end{cases}$$

where

$$(3.96)\qquad \eta_1(y) = 1 + o(1),\ \eta_2(y) = 1 + o(1),\ y \to \infty.$$

Substituting in (3.86) with $c_1 = 0$, $c_2 = 1$ and $y_0 = \infty$ gives, after simplification,

$$(3.97)\qquad \begin{aligned} \eta_1(y) &= 1 - \int_y^\infty \{\rho(\infty)(y'-y^{-1}y'^2)B_2(y')\eta_1(y') + \rho^{-1}(\infty)y^{-1}B_1(y')\eta_2(y')\}dy' \\ \eta_2(y) &= 1 - \int_y^\infty \rho(\infty)y'B_2(y')\eta_1(y')dy'. \end{aligned}$$

or

$$(3.98)\qquad \eta(y) = \eta^0 + \int_{y_1}^\infty K(y,y')\,\eta(y')\,dy',\ y \ge y_1$$

where $\eta(y)$ and η^0 have components $\eta_1(y)$, $\eta_2(y)$ and 1, 1 respectively, $K(y,y') \equiv 0$ for $y \ge y'$ and

$$
\text{(3.99)}\qquad
\begin{aligned}
K_{11}(y,y') &= \rho(\infty)(-y' + y^{-1}y'^2)B_2(y') \\
K_{12}(y,y') &= -\rho^{-1}(\infty)y^{-1}\, B_1(y') \\
K_{21}(y,y') &= -\rho(\infty)\, y'\, B_2(y') \\
K_{22}(y,y') &= 0
\end{aligned}
$$

for $y \le y'$. It follows from (3.91) that

$$
|(K\eta)_1(y)| \le \|\eta\| \left\{\rho(\infty)\int_y^\infty (y'+y^{-1}y'^2)|B_2(y')|dy' + \rho^{-1}(\infty)y^{-1}\int_y^\infty |B_1(y')|dy'\right\} \tag{3.100}
$$

and

$$
|(K\eta)_2(y)| \le \|\eta\|\, \rho(\infty)\int_y^\infty y'\, |B_2(y')|\, dy'. \tag{3.101}
$$

In particular,

$$
\begin{aligned}
\|K\| \le \rho(\infty)\int_{y_1}^\infty y'|B_2(y')|dy' + \rho(\infty)y_1^{-1}\int_{y_1}^\infty y'^2|B_2(y')|dy' \\
+ \rho^{-1}(\infty)y_1^{-1}\int_{y_1}^\infty |B_1(y')|dy' + \rho(\infty)\int_{y_1}^\infty y'|B_2(y')|dy'.
\end{aligned} \tag{3.102}
$$

Hence (3.85) implies that K is contractive in X for y_1 large enough and a unique solution is obtained as in the preceding case. Finally, (3.100) and (3.101) together with (3.85) imply that (3.96) is satisfied.

The existence of the special solutions ϕ_1, ϕ_2 has thus been proved. Their linear independence follows directly from (3.21), (3.22).

Proof of Corollary 3.7. This was verified by (3.25).

§4. SPECTRAL PROPERTIES OF A_μ

The results of §3 are used in this section to derive precise results concerning the location and nature of the spectrum of A_μ. The notations $\sigma(A_\mu)$, $\sigma_0(A_\mu)$, $\sigma_c(A_\mu)$ and $\sigma_e(A_\mu)$ will be used to denote the spectrum, point spectrum, continuous spectrum and essential spectrum of A_μ, respectively. The definitions of [11, Ch. X] will be used. In particular, $\sigma_c(A_\mu)$ is a closed set and $\sigma_e(A_\mu)$ is the set of all non-isolated points of $\sigma(A_\mu)$. Note that the properties of A_μ described by (2.3) imply that $\sigma(A_\mu) \subset [c_m^2\mu^2,\infty)$.

The Point Spectrum of A_μ. Theorem 3.1 and its corollaries imply the following three lemmas concerning $\sigma_0(A_\mu)$.

Lemma 4.1. For all $\mu > 0$,

$$\sigma_0(A_\mu) \subset [c_m^2\mu^2, c^2(\infty)\mu^2]. \tag{4.1}$$

Lemma 4.2. For all $\mu > 0$,

$$\sigma(A_\mu) \cap [c_m^2\mu^2, c^2(\infty)\mu^2) \subset \sigma_0(A_\mu). \tag{4.2}$$

Moreover, $\sigma_0(A_\mu)$ is either a finite set (possibly empty) or a countable set with unique limit point $c^2(\infty)\mu^2$.

Lemma 4.3. The eigenvalues of A_μ that lie in the interval $[c_m^2\mu^2, c^2(\infty)\mu^2)$ are all simple.

The possibility that $c^2(\infty)\mu^2 \in \sigma_0(A_\mu)$ is not excluded by the hypotheses (2.1.4), (1.1) alone. Criteria for $c^2(\infty)\mu^2 \notin \sigma_0(A_\mu)$ are given below.

It will be convenient to use a notation that permits a unified discussion of the cases of finite and infinite point spectra $\sigma_0(A_\mu)$. The number of eigenvalues in $[c_m^2\mu^2, c^2(\infty)\mu^2)$ will be denoted by $N(\mu) - 1$. Thus $N(\mu)$ is an extended integer-valued function of $\mu > 0$ $(1 \leq N(\mu) \leq +\infty)$. The eigenvalues of A_μ in $[c_m^2\mu^2, c^2(\infty)\mu^2)$, arranged in ascending order will be denoted by $\lambda_k(\mu)$, $1 \leq j < N(\mu)$. Thus

$$c_m^2\mu^2 \leq \lambda_1(\mu) < \lambda_2(\mu) < \cdots < c^2(\infty)\mu^2. \tag{4.3}$$

The corresponding eigenfunctions are

$$\psi_k(y,\mu) = a_k(\mu)\ \phi_2(y,\mu,\lambda_k(\mu)),\ k = 1,2,\cdots \tag{4.4}$$

where $a_k(\mu) > 0$ is chosen to make $\|\psi_k(\cdot,\mu)\| = 1$.

The Continuous and Essential Spectra of A_μ. Lemma 4.2 implies that $\sigma_e(A_\mu) \subset [c^2(\infty)\mu^2,\infty)$. Moreover, $\sigma_c(A_\mu)$ and $\sigma_e(A_\mu)$ are closed and $\sigma_c(A_\mu) \subset \sigma_e(A_\mu)$ [11, Ch. X]. The characterization of these sets will be completed in §6 by showing that $(c^2(\infty)\mu^2,\infty) \subset \sigma_c(A_\mu)$. These facts imply

Theorem 4.4. For all $\mu > 0$,

$$\sigma_c(A_\mu) = \sigma_e(A_\mu) = [c^2(\infty)\mu^2,\infty). \tag{4.5}$$

A direct proof of Theorem 4.4 can be given by using the special solutions of §3 and a criterion of Weyl; see [5, p. 1435].

It is known that the bottom point in the essential spectrum of a Sturm-Liouville operator A can be characterized by the oscillation properties of the solutions of $A\phi = \lambda\phi$ [5, p. 1469]. For the operator A_μ the characterization is described by

Corollary 4.5. The equation $A_\mu\phi = \lambda\phi$ is oscillatory (every real solution has infinitely many zeros) for every $\lambda > c^2(\infty)\mu^2$. The equation is non-oscillatory (every real solution has finitely many zeros) for every $\lambda < c^2(\infty)\mu^2$.

These results for A_μ follow directly from Theorem 3.1.

The Point Spectrum of A_μ (continued). The equation $A_\mu\phi = \lambda\phi$ may or may not be oscillatory for $\lambda = c^2(\infty)\mu^2$. This property is shown below to provide a criterion for $\sigma_0(A_\mu)$ to be finite. The basic tool in establishing such criteria is the classical oscillation theorem of Sturm. A version suitable for application to A_μ may be formulated as follows.

Let $I = (a,b)$ be an arbitrary interval $(-\infty \leq a < b \leq +\infty)$ and consider a pair of equations

$$L_j\phi \equiv (P_j^{-1}(y)\phi')' + Q_j(y)\phi = 0, \ j = 1,2, \tag{4.6}$$

where $P_j(y)$ and $Q_j(y)$ are defined and real valued for almost every $y \in I$, $P_j(y) > 0$ for almost every $y \in I$ and P_j, Q_j are Lebesgue integrable on compact subsets of I $(j = 1,2)$. A solution of (4.6) on I is a function $\phi \in AC(I)$ such that $P_j^{-1}\phi' \in AC(I)$ and (4.6) holds for almost all $y \in I$. Such solutions are uniquely determined by the values $\phi(y_0) = c_0$, $P_j^{-1}(y_0)\phi'(y_0) = c_1$ at any point $y_0 \in I$. Pairs of equations (4.6) such that

$$P_1(y) \leq P_2(y), \ Q_1(y) \leq Q_2(y) \text{ for almost all } y \in I \tag{4.7}$$

will be considered. When (4.7) holds the operator L_2 is said to be a Sturm majorant of operator L_1, and the operator L_1 is said to be a Sturm minorant of operator L_2, on I. Sturm's theorem may now be formulated as follows.

Theorem 4.6. Let $\phi_j(y) \not\equiv 0$ be solutions of $L_j\phi_j = 0$ on I $(j = 1,2)$ and assume that y_1 and y_2 are successive zeros of $\phi_1(y)$ in I, with $y_1 < y_2$. Moreover, let L_2 be a Sturm majorant of L_1 on (y_1,y_2). Then $\phi_2(y)$ has at least one zero in $[y_1,y_2)$. In addition, if either $Q_1(y) < Q_2(y)$ or $P_1(y) < P_2(y)$ and $Q_2(y) \neq 0$ on a subset of (y_1,y_2) having positive measure then $\phi_2(y)$ has a zero in (y_1,y_2).

The special solution $\phi_3(y,\mu,\lambda)$ is real valued, tends to zero exponentially when $y \to -\infty$ and has finitely many zeros when $\lambda < c^2(\infty)\mu^2$. Theorem 4.6 will be shown to imply

Corollary 4.7. If $\lambda_1 < \lambda_2 < c^2(\infty)\mu^2$ then $\phi_3(y,\mu,\lambda_2)$ has at least as many zeros as $\phi_3(y,\mu,\lambda_1)$.

It will be convenient following [5, p. 1473] to introduce the sets

(4.8) $$I_k = I_k(\mu) = \{\lambda \mid \phi_3(y,\mu,\lambda) \text{ has exactly } k \text{ zeros}\},\ k = 0,1,2,\cdots.$$

Note that by Corollary 4.5 each $I_k \subset (-\infty, c^2(\infty)\mu^2]$. The point $c^2(\infty)\mu^2$ may or may not be in one of the sets I_k. Corollary 4.7 implies that each I_k is an interval and I_k lies to the left of I_{k+1} for $k = 0,1,2,\cdots$. It is important for the analysis of $\sigma_0(A_\mu)$ to know that the intervals $I_k \neq \phi$ for $k = 1,2,\cdots,N(\mu) - 1$. This is a corollary of the following fundamental oscillation theorem.

Theorem 4.8. If $\sigma_0(A_\mu) \neq \phi$ then for $k = 1,2,\cdots,N(\mu) - 1$ the eigenfunction $\psi_k(y,\mu)$ has precisely $k - 1$ zeros.

Corollary 4.9. If $\sigma_0(A_\mu) \neq \phi$ then

(4.9) $$I_k = (\lambda_k(\mu), \lambda_{k+1}(\mu)],\ k = 0,1,\cdots,N(\mu) - 2$$

(where $\lambda_0(\mu) \equiv -\infty$). Moreover, if $N(\mu) < \infty$ then $(\lambda_{N(\mu)-1}(\mu), c^2(\infty)\mu^2) \subset I_{N(\mu)-1}$.

Corollary 4.10. The number of eigenvalues that satisfy $\lambda_k(\mu) < \lambda < c^2(\infty)\mu^2$ is equal to the number of zeros of $\phi_3(y,\mu,\lambda)$.

Criteria for the Finiteness of $\sigma_0(A_\mu)$. The principal criterion for $\sigma_0(A_\mu)$ to be finite is described by

Theorem 4.11. $\sigma_0(A_\mu)$ is finite if and only if the equation $A_\mu\phi = c^2(\infty)\mu^2\phi$ is non-oscillatory on R. Hence $\sigma_0(A_\mu)$ is infinite if and only if $A_\mu\phi = c^2(\infty)\mu^2\phi$ is oscillatory on R.

It is shown below that Theorem 4.11 is a consequence of Theorem 4.8 and Sturm's comparison theorem.

Corollary 4.12. If $c(\infty) < c(-\infty)$ then $\sigma_0(A_\mu)$ is finite if and only if $\phi_3(y,\mu,c^2(\infty)\mu^2)$ has only a finite number of zeros.

Specific criteria for the finiteness of $\sigma_0(A_\mu)$ will now be obtained by deriving criteria for $A_\mu\phi = c^2(\infty)\mu^2\phi$ to be non-oscillatory and using Theorem 4.11. It will be assumed that $c(\infty) < c(-\infty)$ so that $A_\mu\phi = c^2(\infty)\mu^2\phi$ is non-oscillatory in neighborhoods of $y = -\infty$. Cases for which $c(\infty) = c(-\infty)$ may be treated by applying non-oscillation criteria at both $y = \infty$ and $y = -\infty$.

A criterion for $\sigma_0(A_\mu)$ to be finite is provided by Theorem 3.6. For under the conditions of the theorem $A_\mu\phi = c^2(\infty)\mu^2\phi$ has a solution basis ϕ_1, ϕ_2 satisfying (3.21), (3.22). It follows that the equation is non-oscillatory. This implies

Theorem 4.13. If $\rho(y)$, $c(y)$ satisfy (2.1.4), (1.1), $c(\infty) < c(-\infty)$ and

$$\int_0^\infty y^2 \, |c(y) - c(\infty)| \, dy < \infty \tag{4.10}$$

then $\sigma_0(A_\mu)$ is finite for every $\mu > 0$.

Alternative criteria for the finiteness of $\sigma_0(A_\mu)$ can be derived by constructing Sturm majorants of $A_\mu\phi = c^2(\infty)\mu^2\phi$ that are non-oscillatory and using Theorems 4.6 and 4.11. Similarly, criteria for $\sigma_0(A_\mu)$ to be infinite can be derived by constructing Sturm minorants that are oscillatory. Several criteria of this type will be given.

The equation $A_\mu\phi = \lambda\phi$ can be written

$$(\rho^{-1}(y)\phi')' + \rho^{-1}(y)(\lambda c^{-2}(y) - \mu^2)\phi = 0. \tag{4.11}$$

In particular, for $\lambda = c^2(\infty)\mu^2$ one has

$$(\rho^{-1}(y)\phi')' + \rho^{-1}(y)\mu^2(c^2(\infty)c^{-2}(y) - 1)\phi = 0. \tag{4.12}$$

The first factor $\rho^{-1}(y)$ in (4.12) is unimportant for the oscillation properties of the equation. Replacing it by ρ_M^{-1} gives the majorant

$$\phi'' + \rho_M \, \rho^{-1}(y)\mu^2 \, (c^2(\infty)c^{-2}(y) - 1)\phi = 0. \tag{4.13}$$

Each non-oscillatory Sturm majorant of (4.13) gives a criterion for the finiteness of $\sigma_0(A_\mu)$. Since solutions to (4.13) are non-oscillatory on any interval $(-\infty, y_0)$ when $c(\infty) < c(-\infty)$, it is enough to construct majorants of (4.13) on intervals (y_0, ∞). An obvious non-oscillatory majorant for (4.13) is $\phi'' = 0$. Thus $\sigma_0(A_\mu)$ is finite for every $\mu > 0$ if there is a y_0 such that $c^2(\infty)c^{-2}(y) - 1 \le 0$ for all $y \ge y_0$; that is,

$$c(y) \ge c(\infty) \text{ for } y \ge y_0. \tag{4.14}$$

This means the graph of $c = c(y)$ lies above or on the limit line $c = c(\infty)$ in a neighborhood of $y = \infty$. Weaker hypotheses that include this case can be derived by comparing (4.13) with

$$\phi'' + \alpha \, y^{-2} \, \phi = 0 \tag{4.15}$$

which is oscillatory on (y_0, ∞) if $\alpha > 1/4$ and non-oscillatory if $\alpha \le 1/4$.

Oscillation theorems based on (4.15) were first given by A. Kneser [12]; see [5, p. 1463]. Comparison of (4.12) with (4.15) gives

Theorem 4.14. If $c(\infty) < c(-\infty)$ and

$$\limsup_{y\to\infty} y^2(c^{-2}(y) - c^{-2}(\infty)) \leq 0 \tag{4.16}$$

then $\sigma_0(A_\mu)$ if finite for all $\mu > 0$. Conversely, if

$$\liminf_{y\to\infty} y^2(c^{-2}(y) - c^{-2}(\infty)) > 0 \tag{4.17}$$

then there exists a $\mu_0 > 0$ such that $\sigma_0(A_\mu)$ is infinite for every $\mu \geq \mu_0$.

Note that the criterion (4.16) includes (4.14) as a special case. Note also that sufficient conditions for (4.16) or (4.17) to hold are the existence of constants y_0, K and $\varepsilon > 0$ such that

$$c(y) \geq c(\infty) - K\, y^{-2-\varepsilon} \text{ for } y \geq y_0 \tag{4.18}$$

or

$$c(y) \leq c(\infty) - K\, y^{-2} \text{ for } y \geq y_0, \tag{4.19}$$

respectively. In particular, $\sigma_0(A_\mu)$ is finite for all $\mu > 0$ if $c(y)$ approaches $c(\infty)$ from below sufficiently rapidly.

Criteria that Guarantee $\sigma_0(A_\mu) \neq \phi$. Such criteria may be derived by constructing Sturm minorants for $A_\mu\phi = c^2(\infty)\mu^2\phi$ whose solutions have zeros. If the minorant has solutions with infinitely many zeros then $\sigma_0(A_\mu)$ is infinite. If the minorant has a solution with finitely many zeros then it can be shown that $\phi_3(y,\mu,c^2(\infty)\mu^2)$ has at least as many zeros and one may use the following refinement of Theorem 4.11.

Theorem 4.15. If $A_\mu\phi = c^2(\infty)\mu^2\phi$ has a solution having a finite number k of zeros on R then the part of $\sigma(A_\mu)$ below $c^2(\infty)\mu^2$ is finite and has at least k - 1 and at most k + 2 points.

To apply the method in cases where $\sigma_0(A_\mu)$ is finite consider first the case $\rho(y) = \text{const.}$ so that (4.12) becomes

$$\phi'' + \mu^2(c^2(\infty)c^{-2}(y) - 1)\phi = 0. \tag{4.20}$$

Note that $c^2(\infty)c^{-2}(y) - 1 \geq c^2(\infty)c_0^{-2}(y) - 1$ for all $y \in R$ if and only if

$$c(y) \leq c_0(y) \text{ for all } y \in R. \tag{4.21}$$

If $c_0(y)$ can be chosen in such a way that

$$(4.22) \qquad \phi'' + \mu^2(c^2(\infty)c_0^{-2}(y) - 1)\phi = 0$$

has a solution on R with k zeros then $\sigma_0(A_\mu)$ will have at least k - 1 points by Theorem 4.15. In this way one can prove

<u>Theorem 4.16</u>. Let $\rho(y)$ = const. for all $y \in R$ and assume that there is a constant $c_0 \geq c_m$ and an interval $I = [a,b]$ with $b > a$ such that

$$(4.23) \qquad c(y) \leq c_0 < c(\infty) \leq c(-\infty) \text{ for all } y \in I.$$

Then $\sigma_0(A_\mu) \neq \phi$ for all sufficiently large μ. In fact, $N(\mu) \to \infty$ when $\mu \to \infty$.

Theorem 4.16 can be proved by comparing c(y) with a suitable piece-wise constant function $c_0(y)$ that satisfies (4.21). An analogue of Theorem 4.16 can be proved in the general case where $\rho(y) \neq$ const. by making the change of variable $y \to \eta$ in (4.12), where

$$(4.24) \qquad \eta = \int_0^y \rho(y')\, dy'.$$

The details, which are elementary but lengthy, are omitted.

This completes the formulation of the results of §4 and the proofs will now be given. Note that Lemma 4.1 is an immediate consequence of Theorem 3.1 which implies that for $\lambda > c^2(\infty)\mu^2$ the equation $A_\mu\phi = \lambda\phi$ has no solutions in $\mathcal{H}(R)$.

<u>Proof of Lemma 4.2</u>. The resolvent of A_μ is an integral operator in $\mathcal{H}(R)$ [5, XIII.3]:

$$(4.25) \qquad (A_\mu - \zeta)^{-1} f(y) = \int_R G_\mu(y,y',\zeta)\, f(y')\, c^{-2}(y')\, \rho^{-1}(y')\, dy'.$$

$G_\mu(y,y',\zeta)$, the Green's function of A_μ, is known to have the form [5, p. 1329]

$$(4.26) \qquad G_\mu(y,y',\zeta) = [\phi_\infty \phi_{-\infty}]^{-1} \begin{cases} \phi_{-\infty}(y)\, \phi_\infty(y'), & y \leq y', \\ \phi_\infty(y)\, \phi_{-\infty}(y'), & y \geq y', \end{cases}$$

where ϕ_∞ and $\phi_{-\infty}$ are non-trivial solutions of $A_\mu\phi = \zeta\phi$ that are in $L_2(0,\infty)$ and $L_2(-\infty,0)$, respectively. Thus for $\zeta \in L(c(\infty)\mu) \subset L(c(-\infty)\mu)$, $\phi_\infty = \phi_2$, $\phi_{-\infty} = \phi_3$ and one has

$$(4.27)\qquad G_\mu(y,y',\zeta) = [\phi_2\phi_3]^{-1}\begin{cases}\phi_3(y,\mu,\zeta)\ \phi_2(y',\mu,\zeta), & y \le y',\\ \phi_2(y,\mu,\zeta)\ \phi_3(y',\mu,\zeta), & y \ge y'.\end{cases}$$

It follows from Corollary 3.3 that $G_\mu(y,y',\zeta)$ is meromorphic in $L(c(\infty)\mu)$ with poles at the zeros of

$$(4.28)\qquad F(\mu,\zeta) = [\phi_2(\cdot,\mu,\zeta)\ \phi_3(\cdot,\mu,\zeta)].$$

As remarked in §1, these are precisely the eigenvalues of A_μ that are less than $c^2(\infty)\mu^2$. Their only possible limit point is $c^2(\infty)\mu^2$ since $F(\mu,\zeta)$ is analytic in $L(c(\infty)\mu)$ by Corollary 3.3. These results imply the two statements of Lemma 4.2.

Proof of Lemma 4.3. This follows from Theorem 3.1 which implies that $A_\mu\phi = \lambda\phi$ always has at least one solution that is not in $\mathcal{H}(R)$.

Proof of Theorem 4.4. It was remarked above that $\sigma_c(A_\mu) \subset \sigma_e(A_\mu) \subset [c^2(\infty)\mu^2,\infty)$. Hence to prove (4.5) it is enough to show that $(c^2(\infty)\mu^2,\infty) \subset \sigma_c(A_\mu)$. This may be done by constructing a characteristic sequence for each $\lambda \in (c^2(\infty)\mu^2,\infty)$; i.e., a bounded sequence $\{\phi_n(y)\}$ in $\mathcal{H}(R)$ such that each $\phi_n \in D(A_\mu)$ and $(A_\mu - \lambda)\phi_n \to 0$ in $\mathcal{H}(R)$ but $\{\phi_n\}$ has no convergent subsequences. Indeed, a suitable sequence has the form $\phi_n(y) = \xi_n(y)\ \phi_3(y,\mu,\lambda)$ where $\xi_n \in D(A_0)$, $\xi_n(y) \equiv 1$ for $|y| \le n$, $\operatorname{supp} \xi_n \subset [-n-1,n+1]$ and $\xi_n'(y)$ and $(\rho^{-1}(y)\xi_n'(y))'$ are bounded for all y and n. Such a sequence $\{\xi_n\}$ can be constructed but the details are lengthy. They will not be given here since the inclusion $(c^2(\infty)\mu^2,\infty) \subset \sigma_c(A_\mu)$ is proved in §6.

Proof of Corollary 4.5. For $\lambda \ne c^2(\infty)\mu^2$ every solution of $A_\mu\phi = \lambda\phi$ is a linear combination of $\phi_1(y,\mu,\lambda)$ and $\phi_2(y,\mu,\lambda)$ (Corollary 3.7). It follows from Theorem 3.1 that every real solution with $\lambda > c^2(\infty)\mu^2$ has infinitely many zeros in any interval (y_0,∞). On the other hand for $\lambda < c^2(\infty)\mu^2$ Theorem 3.1 implies that every real solution of $A_\mu\phi = \lambda\phi$ is either exponentially large or exponentially small for $y \to \pm\infty$. In every case $\phi(y)$ has constant sign outside of some interval $[-y_0,y_0]$ and hence can have only finitely many zeros.

Proof of Theorem 4.6. Results equivalent to Theorem 4.6 are proved in [9, Ch. XI] under the additional hypothesis that the P_j and Q_j are continuous. The same method will be shown to be applicable under the hypotheses of Theorem 4.6. The method is to study the phase plane curves

$$(4.29)\qquad (\xi,\eta) = (P_j^{-1}(y)\phi_j'(y),\phi_j(y)),\ y \in I,\ j = 1,2,$$

defined by the solutions $\phi_j(y)$ and to transform to polar coordinates (Prüfer transformation). Thus (4.29) can be written

$$(4.30)\qquad (\xi,\eta) = (r_j(y)\cos\theta_j(y), r_j(y)\sin\theta_j(y)),\ y \in I,\ j = 1,2.$$

Moreover, the curves (4.29) cannot pass through the origin because $\phi_j(y) \not\equiv 0$. Thus $r_j(y) > 0$ and $\theta_j(y)$ is uniquely defined by continuity and its value at the point $y_1 \in I$. Finally, $\theta_j \in AC(I)$ and (4.6) implies that θ_j is a solution of the first order equation

$$(4.31)\qquad \theta_j'(y) = P_j(y)\cos^2\theta_j(y) + Q_j(y)\sin^2\theta_j(y),\ y \in I.$$

To prove the first statement of Theorem 4.6 note that one can assume without loss of generality that $\phi_1(y) > 0$ for $y_1 < y < y_2$ and $\phi_2(y_1) \geq 0$. Thus $\theta_j(y)$ $(j = 1,2)$ may be defined as the unique solutions of (4.31) such that $\theta_1(y_1) = 0$ and $0 \leq \theta_2(y_1) < \pi$. It follows that $0 < \theta_1(y) < \pi$ for $y_1 < y < y_2$ and $\theta_1(y_2) = \pi$. It must be shown that $\phi_2(y)$ has a zero in $[y_1,y_2)$. If $\phi_2(y_1) = 0$ there is nothing to prove. If $\phi_2(y_1) > 0$ then $0 < \theta_2(y_1) < \pi$ and it follows from (4.31) and (4.7) that $\theta_2(y) > \theta_1(y)$ for all $y \geq y_1$ (see [9, p. 335]). In particular, $\theta_2(y_2) > \theta_1(y_2) = \pi$ whence by continuity $\theta_2(y_0) = \pi$ and therefore $\phi_2(y_0) = 0$ for some $y_0 \in (y_1,y_2)$.

To prove the second statement of theorem 4.6 it is only necessary to remark that if $Q_1(y) < Q_2(y)$ or $P_1(y) < P_2(y)$ and $Q_2(y) \neq 0$ on a subset of (y_1,y_2) having positive measure then $\theta_2(y_2) > \pi$ even if $\theta_2(y_1) = 0$; see [9, p. 335]. This completes the proof.

Proof of Corollary 4.7. The function $\phi_3(y,\mu,\lambda)$ is a solution of equation (4.11). Thus if $\phi_j(y) = \phi_3(y,\mu,\lambda_j)$, $j = 1,2$, then

$$(4.32)\qquad (\rho^{-1}(y)\phi_j')' + \rho^{-1}(y)(\lambda_j c^{-2}(y) - \mu^2)\phi_j = 0,\ j = 1,2.$$

These equations have the form (4.6) with $P_j(y) = \rho(y)$ and $Q_j(y) = \rho^{-1}(y)(\lambda_j c^{-2}(y) - \mu^2)$. Hence $P_1(y) = P_2(y)$ and $Q_1(y) < Q_2(y)$ for all $y \in R$, since $\rho(y)$ and $c(y)$ are always positive, and the second part of Theorem 4.6 is applicable. It follows that if $y_1 < y_2 < \cdots < y_k$ are the zeros of $\phi_1(y) = \phi_3(y,\mu,\lambda_1)$ then $\phi_2(y) = \phi_3(y,\mu,\lambda_2)$ has $k - 1$ zeros in the interval (y_1,y_k). Hence it will be enough to show that $\phi_3(y,\mu,\lambda_2)$ also has a zero in $(-\infty,y_1]$. To verify this apply Lagrange's formula (3.23) to $\phi_1(y)$

and $\phi_2(y)$ in $(-\infty,y_1]$. This is possible because $\phi_1(y)$ and $\phi_2(y)$ are exponentially small at $y = -\infty$. The result can be written

$$(\lambda_2 - \lambda_1) \int_{-\infty}^{y_1} \phi_1(y)\, \phi_2(y)\, c^{-2}(y)\, \rho^{-1}(y)dy$$

$$(4.33) \qquad = \int_{-\infty}^{y_1} \{\phi_1\, A_\mu\phi_2 - \phi_2\, A_\mu\phi_1\}\, c^{-2}\, \rho^{-1}\, dy$$

$$= \phi_2(y_1)\, \{\rho^{-1}(y_1)\, \phi_1'(y_1)\}$$

since $\phi_1(y_1) = 0$ and ϕ_1 and ϕ_2 vanish at $y = -\infty$. Now suppose that $\phi_2(y)$ has no zero in $(-\infty,y_1]$. Then $\phi_2(y_1) > 0$ because $\phi_2(y) > 0$ near $y = -\infty$ by Theorem 3.1. Moreover, $\phi_1(y) > 0$ for $-\infty < y < y_1$ and $\rho^{-1}(y_1)\, \phi_1'(y_1) < 0$ because y_1 is the first zero of ϕ_1. Thus the right hand side of (4.33) is negative. But the left hand side is clearly positive. This contradiction completes the proof.

Proof of Theorem 4.8. For regular Sturm-Liouville operators $A\phi = -\phi'' + q(y)\phi$ on finite intervals the oscillation theorem goes back to Sturm. For singular operators in $L_2(0,\infty)$ such that $q(y) \to +\infty$ when $y \to +\infty$ it was first proved by H. Weyl. More recently the result was proved by B. M. Levitan and I. S. Sargsjan [14, p. 201] under more general conditions on $Q(y)$ that guarantee that the solutions of $A\phi = \lambda\phi$ are non-oscillatory on $0 \le y < \infty$ for all $\lambda \in R$ (and hence $\sigma(A)$ is discrete). It will be shown here that the method of Levitan and Sargsjan is applicable to the case of Theorem 4.8.

The method of [14] is to regard the Sturm-Liouville problem for A on $0 \le y < \infty$ as a limit of regular problems for A on $0 \le y \le b < \infty$ and to study the behavior of the eigenvalues and eigenfunctions as $b \to \infty$. Here the operator A_μ in $\mathcal{H}(R)$ will be regarded as a limit of the regular operator in $\mathcal{H}(a,b) = L_2(a,b;c^{-2}(y)\, \rho^{-1}(y)dy)$, $-\infty < a < b < \infty$, defined by A_μ and the boundary conditions $\phi(a) = \phi(b) = 0$. The corresponding operator in $\mathcal{H}(a,b)$ will be denoted by $A_{\mu,a,b}$. The limit $a \to -\infty$ will be studied first.

The operator A_μ is more general than the operator studied in [14]. However, the examination of the proofs in [14] reveals that nothing is used but the Sturm comparison theorem, the convergence of the eigenvalues when $b \to \infty$, the continuity and asymptotic properties of the solution $\phi(y,\lambda)$ of $A\phi = \lambda\phi$ that satisfies the boundary condition at $y = 0$ and the non-oscillatory character of $A\phi = \lambda\phi$ in a λ-interval containing the point spectrum. All of these properties have been established for A_μ.

The solution of $A_\mu\phi = \lambda\phi$ that satisfies $\phi(b) = 0$, $\rho^{-1}(b)\,\phi'(b) = 1$ will be denoted by $\phi_b(y,\lambda)$. For $\lambda \neq c^2(-\infty)\mu^2$, $\phi_b(y,\lambda)$ is a linear combination of $\phi_3(y,\mu,\lambda)$ and $\phi_4(y,\mu,\lambda)$ and hence has the regularity properties of Corollaries 3.2 and 3.3. The eigenvalues of $A_{\mu,a,b}$ are the roots of the equation $\phi_b(a,\lambda) = 0$. They will be denoted by $\lambda_{k,a,b}$, $k = 1,2,\cdots$, with the convention that $\lambda_{k,a,b} < \lambda_{k+1,a,b}$. The corresponding eigenfunctions $\psi_{k,a,b}(y) = \phi_b(y,\lambda_{k,a,b})$ have precisely $k - 1$ zeros by the classical oscillation theorem. For the class of operators considered here this result can be proved by the method of [14, p. 17]. The zeros of $\psi_{k,a,b}(y)$ will be denoted by $y^{(k)}_{j,a,b}$, $1 \leq j \leq k - 1$.

The operator in $\mathcal{H}(-\infty,b)$ defined by A_μ and the boundary condition $\phi(b) = 0$ will be denoted by $A_{\mu,b}$. The methods used to study $\sigma(A_\mu)$ above can be used to show that $\sigma(A_{\mu,b}) \cap [c_m^2\mu^2, c^2(-\infty)\mu^2)$ $(\subset \sigma_0(A_{\mu,b}))$ is finite or countably infinite with unique limit point $c^2(-\infty)\mu^2$. The number of eigenvalues in $[c_m^2\mu^2, c^2(-\infty)\mu^2)$ will be denoted by $N(\mu,b) - 1$ $(\leq +\infty)$ in analogy with the notation for A_μ, and the eigenvalues will be denoted by $\lambda_{k,b}$ $(\lambda_{k,b} < \lambda_{k+1,b})$. The eigenfunctions for $A_{\mu,b}$ are $\psi_{k,b}(y) = \phi_b(y,\lambda_{k,b})$.

The proof of the oscillation theorem for $A_{\mu,b}$ by the method of [14] will now be outlined. First

$$\lim_{a\to-\infty} \lambda_{k,a,b} = \lambda_{k,b} \text{ for } 1 \leq k < N(\mu,b). \tag{4.34}$$

This follows, for example, from the convergence of the Green's functions. It follows that

$$\lim_{a\to-\infty} \psi_{k,a,b}(y) = \psi_{k,b}(y) \text{ for } -\infty < y \leq b, \tag{4.35}$$

uniformly on bounded subsets of $(-\infty,b]$. The proof of the oscillation theorem given in [14] is based on the following three lemmas.

<u>Lemma 4.17</u>. For each $k = 1,2,\cdots,N(\mu,b) - 1$ and each fixed $b \in R$ one has

$$\sup_{a<b} y^{(k)}_{k-1,a,b} < \infty. \tag{4.36}$$

<u>Lemma 4.18</u>. For each $k = 1,2,\cdots,N(\mu,b) - 1$, each $j = 1,2,\cdots,k - 1$ and each fixed $b \in R$ one has

$$\inf_{a<b} \left(y^{(k)}_{j+1,a,b} - y^{(k)}_{j,a,b} \right) > 0. \tag{4.37}$$

Lemma 4.19. For each $k = 1,2,\cdots,N(\mu,b) - 1$ and each fixed $b \in R$ one has

$$\inf_{a<b} (\lambda_{k+1,a,b} - \lambda_{k,a,b}) > 0. \tag{4.38}$$

The proofs of these lemmas and the oscillation theorem for $A_{\mu,b}$ are the same as those given in [14, pp. 202-4] and will not be repeated here.

The proof of Theorem 4.8 may now be completed by regarding A_μ as a limit of $A_{\mu,b}$ for $b \to \infty$ and repeating the argument given above. The solution of $A_\mu\phi = \lambda\phi$ that satisfies the condition of square integrability at $y = -\infty$ is $\phi_3(y,\mu,\lambda)$ and is non-oscillatory for all $y \in R$ when $\lambda < c^2(\infty)\mu^2$ (Corollary 4.5). The remainder of the proof follows as before.

Proof of Corollary 4.9. Theorem 4.8 implies that $\lambda_{k+1}(\mu) \in I_k$ for $k = 0,1,2,\cdots,N(\mu) - 2$. Moreover, a continuity argument based on Corollary 3.2 shows that the intervals I_k have the form $I_k = (a_k,a_{k+1}]$ where $a_0 < a_1 < \cdots < a_{N(\mu)-1}$ (see [5, p. 1475]). Thus to prove (4.9) it will suffice to prove that

$$F(\mu,a_k) = [\phi_2(\cdot,\mu,a_k)\ \phi_3(\cdot,\mu,a_k)] = 0 \tag{4.39}$$

for $k = 1,2,\cdots,N(\mu) - 1$. This proof will be based on the following two lemmas.

Lemma 4.20. Let $\lambda_0 < c^2(-\infty)\mu^2$ and let y_0 be a zero of $\phi_3(y,\mu,\lambda_0)$. Then to each sufficiently small $\varepsilon > 0$ there corresponds a $\delta > 0$ such that for $|\lambda - \lambda_0| < \delta$ the function $\phi_3(y,\mu,\lambda)$ has exactly one zero in the interval $|y - y_0| < \varepsilon$.

This result follows from Corollary 3.2 and the fact that $\rho^{-1}(y)\ \phi'(y)$ cannot vanish at a zero of a non-trivial solution of $A_\mu\phi = \lambda\phi$. For a proof see [14, p. 16].

For $\lambda \in I_k$ let $y_1(\lambda) < y_2(\lambda) < \cdots < y_k(\lambda)$ denote the zeros of $\phi_3(y,\mu,\lambda)$. Then each $y_j(\lambda)$ is uniquely defined for $\lambda \in \cup_{k\geq j} I_k$ and one has

Lemma 4.21. Each of the functions $y_j(\lambda)$ is continuous and strictly monotone increasing.

The continuity follows immediately from Lemma 4.20. The strict monotonicity follows from the proof of Corollary 4.7.

Proof of Corollary 4.9 (concluded). (4.39) will be proved by contradiction. Assume that $F(\mu,a_k) \neq 0$ and note that for $\lambda < c^2(\infty)\mu^2$ one has

$$\phi_3(y,\mu,\lambda) = c(\mu,\lambda)\ \phi_1(y,\mu,\lambda) + c'(\mu,\lambda)\ \phi_2(y,\mu,\lambda), \tag{4.40}$$

by Corollary 3.7. Moreover, Theorem 3.1 implies that

(4.41) $$c(\mu,\lambda) = \rho(\infty)[\phi_2\phi_3]/2iq_+(\mu,\lambda) = \rho(\infty)\ F(\mu,\lambda)/2iq_+(\mu,\lambda).$$

Thus $c(\mu,a_k) \neq 0$ and by continuity (Corollary 3.2) there is an interval $|\lambda - a_k| \leq \delta$ in which $c(\mu,\lambda) \neq 0$. It follows from (4.40) and the uniformity of the asymptotic estimates of Theorem 3.1 (Corollary 3.4) that there is an $M > 0$ such that

(4.42) $$|\phi_3(y,\mu,\lambda)| \geq 1 \text{ for all } y \geq M \text{ and } |\lambda - a_k| \leq \delta.$$

Note that Lemma 4.21 implies

(4.43) $$\lim_{\lambda\to a_k} y_j(\lambda) = y_j(a_k),\ j = 1,2,\cdots,k.$$

Now consider $y_{k+1}(\lambda)$ which is defined for $\lambda > a_k$. For $a_k < \lambda < a_k + \delta$ one has $y_{k+1}(\lambda) < M$ by (4.42). It follows that the limit

(4.44) $$\bar{y} = \lim_{\lambda\to a_k} y_{k+}(\lambda)$$

exists and $\bar{y} \geq y_k(a_k)$. But $\phi_3(\bar{y},\mu,a_k) = 0$ by continuity and hence $\bar{y} = y_k(a_k)$. But this implies that for $\delta > 0$ small enough and $a_k < \lambda < a_k + \delta$ every neighborhood $|y - y_k(a_k)| < \varepsilon$ contains two zeros of $\phi_3(y,\mu,\lambda)$ in contradiction to Lemma 4.20. This completes the proof of (4.9).

The last statement of Corollary 4.9 follows from the proof of Theorem 4.8. Indeed, if $N(\mu) < \infty$ and $N(\mu) < N(\mu,b)$ then $N(\mu) \leq N(\mu,b) - 1$ and $\lambda_{N(\mu),b}$ is defined. In this case

(4.45) $$\liminf_{b\to\infty} \lambda_{N(\mu),b} \geq c^2(\infty)\mu^2$$

since otherwise $\lambda_{N(\mu),b(k)} \to \lambda_0 < c^2(\infty)\mu^2$ for some subsequence $\{b(k)\}$, which would imply that λ_0 was an additional eigenvalue of A_μ. If $N(\mu) = N(\mu,b) < \infty$ the same argument can be applied to the operator $A_{\mu,b}$.

Proof of Corollary 4.10. This follows immediately from Corollary 4.9.

Proof of Theorem 4.11. It will suffice to prove the second statement of the theorem. To this end let $\phi_b(y,\mu,\lambda)$ be the solution of $A_\mu\phi = \lambda\phi$ that satisfies $\phi_b(b,\mu,\lambda) = 0$, $\rho^{-1}(b)\ \phi_b'(b,\mu,\lambda) = 1$. Then $\phi_b(y,\mu,\lambda)$ and $\rho^{-1}(y)\ \phi_b'(y,\mu,\lambda)$ are continuous functions of $(y,\lambda) \in R^2$. Now assume that $A_\mu\phi = c^2(\infty)\mu^2\phi$ is oscillatory. Then $\phi_b(y,\mu,c^2(\infty)\mu^2)$ has infinitely many zeros. It follows by the method used to prove Lemma 4.20 that the number of zeros of $\phi_b(y,\mu,\lambda)$ tends to infinity as $\lambda \to c^2(\infty)\mu^2$. But then the same is true of $\phi_3(y,\mu,\lambda)$, by Theorem 4.6, and it follows from Corollary 4.10 that $\sigma_0(A_\mu)$ is infinite. To prove the converse note that if $\sigma_0(A_\mu)$ is infinite then Theorem 4.6, applied to the kth eigenfunction and any solution of $A_\mu\phi = c^2(\infty)\mu^2\phi$ implies that ϕ has k - 2 zeros. Since k is arbitrary it follows that $A_\mu\phi = c^2(\infty)\mu^2\phi$ is oscillatory.

Proof of Corollary 4.12. This follows immediately from Theorem 4.11. The hypothesis $c(\infty) < c(-\infty)$ is needed only to ensure that $\phi_3(y,\mu,c^2(\infty)\mu^2)$ is defined.

Proof of Theorem 4.13. This follows immediately from Corollary 4.12 and Theorem 3.6.

Proof of Theorem 4.14. To prove the first half of the theorem it will be shown that condition (4.16) implies the existence of a non-oscillatory majorant for equation (4.13) for every $\mu > 0$. This implies that (4.12), i.e., $A_\mu\phi = c^2(\infty)\mu^2\phi$, is non-oscillatory for every $\mu > 0$ and the finiteness of $\sigma_0(A_\mu)$ follows from Corollary 4.12.

To construct a majorant for (4.13) note that (4.16) implies that for every $\varepsilon > 0$ there is a $y_0 = y_0(\varepsilon)$ such that

$$y^2(c^2(\infty)c^{-2}(y) - 1)_+ \leq \varepsilon \text{ for all } y \geq y_0(\varepsilon), \tag{4.46}$$

where $\alpha_+ = \text{Max}(\alpha,0)$. It follows that for every $\mu > 0$ there is a $y_0 = y_0(\mu)$ such that

$$\begin{aligned} \rho_M\, \rho^{-1}(y)\, y^2(c^2(\infty)c^{-2}(y) - 1) &\leq \rho_M\, \rho^{-1}(y)\, y^2(c^2(\infty)c^{-2}(y) - 1)_+ \\ &\leq \rho_M\, \rho_m^{-1}\, y^2(c^2(\infty)c^{-2}(y) - 1)_+ \\ &\leq \frac{1}{4}\mu^2 \text{ for all } y \geq y_0(\mu). \end{aligned} \tag{4.47}$$

Hence for any $\mu > 0$ one has

$$\rho_M\, \rho^{-1}(y)\, \mu^2(c^2(\infty)c^{-2}(y) - 1) \leq \frac{1}{4} y^2 \text{ for all } y \geq y_0(\mu). \tag{4.48}$$

It follows on comparing (4.13) with (4.15) with $\alpha = 1/4$ that (4.13) is non-oscillatory on $y_0(\mu) \leq y < \infty$. It is non-oscillatory on $-\infty < y \leq y_0(\mu)$ for any $\mu > 0$ because $c(\infty) < c(-\infty)$. This proves the first half of Theorem 4.14.

To prove the second half it will be shown that (4.17) implies the existence of a $\mu_0 > 0$ such that $A_\mu\phi = c^2(\infty)\mu^2\phi$ is oscillatory for every $\mu \geq \mu_0$. The result then follows from Theorem 4.11. To this end note that if ε satisfies

$$(4.49) \qquad 0 < \varepsilon < \liminf_{y\to\infty} y^2(c^2(\infty)c^{-2}(y) - 1)$$

then there is a $y_0 = y_0(\varepsilon)$ such that

$$(4.50) \qquad y^2(c^2(\infty)c^{-2}(y) - 1) \geq \varepsilon \text{ for all } y \geq y_0(\varepsilon).$$

In particular, given any $\alpha > 1/4$ there is a $\mu_0 > 0$ such that

$$(4.51) \qquad 0 < \rho_M\, \rho_m^{-1}\, \alpha/\mu_0^2 < \liminf_{y\to\infty} y^2(c^2(\infty)c^{-2}(y) - 1).$$

It follows that there is a $y_0 = y_0(\alpha)$ such that

$$(4.52) \qquad y^2(c^2(\infty)c^{-2}(y) - 1) \geq \rho_M\, \rho_m^{-1}\, \alpha/\mu_0^2 \text{ for all } y \geq y_0(\alpha).$$

This implies that

$$(4.53) \qquad \rho_m\, \rho_M^{-1}\, \mu_0^2(c^2(\infty)c^{-2}(y) - 1) \geq \alpha/y^2 \text{ for all } y \geq y_0(\alpha).$$

Hence, comparison of

$$(4.54) \qquad \phi'' + \rho_m\, \rho_M^{-1}\, \mu_0^2(c^2(\infty)c^{-2}(y) - 1)\phi = 0$$

and (4.15) with $\alpha > 1/4$ implies that (4.54) is oscillatory. But (4.54) is a Sturm minorant of (4.12); i.e., $A_\mu\phi = c^2(\infty)\mu^2\phi$, provided $\mu \geq \mu_0$. Hence the latter is oscillatory for all $\mu \geq \mu_0$.

Proof of Theorem 4.15. This result is proved in [5, p. 1481] for Sturm-Liouville operators with smooth coefficients. The proof is based on the oscillation theorem (Theorem 4.8), Sturm's comparison theorem and the continuous dependence of the zeros of solutions of $A_\mu\phi = \lambda\phi$ on λ (Lemma 4.20). Hence it extends immediately to the operator A_μ.

§5. GENERALIZED EIGENFUNCTIONS OF A_μ

The eigenfunctions $\psi_k(y,\mu)$ corresponding to the point spectrum of A_μ were constructed in the preceding section. In this section the special solutions ϕ_j ($j = 1,2,3,4$) of §3 are used to construct generalized eigenfunctions of A_μ corresponding to the points of the continuous spectrum. These functions will be used in §6 to construct the spectral family $\{\Pi_\mu(\lambda)\}$ of A_μ and to prove that $\sigma_c(A_\mu) = \sigma_e(A_\mu) = [c^2(\infty)\mu^2,\infty)$.

To construct the generalized eigenfunctions $\psi_0(y,\mu,\lambda)$, $\psi_\pm(y,\mu,\lambda)$ described in §1 recall that the special solutions $\phi_j(y,\mu,\lambda)$ are defined for all real $\lambda \neq c^2(\pm\infty)\mu^2$ and the pairs ϕ_1, ϕ_2 and ϕ_3, ϕ_4 are solution bases for $A_\mu\phi = \lambda\phi$ (Corollary 3.7). It follows that

$$(5.1)\qquad \begin{cases} \phi_j = c_{j3}\phi_3 + c_{j4}\phi_4, \; j = 1,2, \\ \phi_j = c_{j1}\phi_1 + c_{j2}\phi_2, \; j = 3,4. \end{cases}$$

The coefficients $c_{jk} = c_{jk}(\mu,\lambda)$ can be calculated by means of the bracket operation

$$(5.2)\qquad [\phi_j\phi_k](\mu,\lambda) = [\phi_j(\cdot,\mu,\lambda)\;\phi_k(\cdot,\mu,\lambda)]$$

of Lagrange's formula. Indeed, by forming the brackets of equations (5.1) with ϕ_4, ϕ_3, ϕ_2 and ϕ_1 in succession and using the asymptotic forms of Theorem 3.1 one finds

$$(5.3)\qquad \left.\begin{aligned} (-2iq_-)c_{j3} &= \rho(-\infty)[\phi_j\phi_4] \\ (2iq_-)c_{j4} &= \rho(-\infty)[\phi_j\phi_3] \end{aligned}\right\} \; j = 1,2,$$

$$\left.\begin{aligned} (-2iq_+)c_{j1} &= \rho(\infty)[\phi_j\phi_2] \\ (2iq_+)c_{j2} &= \rho(\infty)[\phi_j\phi_1] \end{aligned}\right\} \; j = 3,4.$$

In particular, Corollary 3.2 implies that each $c_{jk}(\mu,\lambda)$ is a continuous function for $\lambda \neq c^2(\pm\infty)\mu^2$. These relations will be used to determine the generalized eigenfunctions of A_μ. The notation

$$(5.4)\qquad \Lambda = \Lambda(\mu) = \{\lambda \mid c^2(-\infty)\mu^2 < \lambda\},$$

(5.4 cont.) $$\Lambda_0 = \Lambda_0(\mu) = \{\lambda \mid c^2(\infty)\mu^2 < \lambda < c^2(-\infty)\mu^2\}$$

will be used. Note that $\Lambda_0 \neq \phi$ only if $c(\infty) < c(-\infty)$.

The Spectral Interval Λ. The generalized eigenfunctions of A_μ are the bounded solutions of the differential equation $A_\mu\phi = \lambda\phi$. For $\lambda \in \Lambda$, Theorem 3.1 and the relations (5.1) imply that all the solutions are bounded. It will be shown that the functions

$$(5.5) \qquad \begin{cases} \psi_+(y,\mu,\lambda) = a_+(\mu,\lambda)\ \phi_4(y,\mu,\lambda) \\ \psi_-(y,\mu,\lambda) = a_-(\mu,\lambda)\ \phi_1(y,\mu,\lambda) \end{cases}$$

have the asymptotic forms described in §1. The completeness in $\Pi_\mu(\Lambda)\ \mathcal{H}(R)$ of these functions will be proved in §6. The pair (ϕ_2,ϕ_3), which provides an alternative basis, will not be treated explicitly here. It may be shown to correspond to the second family $\{\phi_-(y,p,q) \mid q > 0\}$ described at the end of §1.

It follows from (5.5), (5.1) and Theorem 3.1 that the asymptotic behavior of ψ_+ is given by

$$(5.6) \qquad \psi_+(y,\mu,\lambda) = a_+ \begin{cases} c_{41}\ e^{iq_+y} + c_{42}\ e^{-iq_+y} + o(1), & y \to +\infty, \\ e^{-iq_-y} + o(1), & y \to -\infty, \end{cases}$$

The equivalence of (5.6) and the asymptotic form (1.19) of §1 follows from

Lemma 5.1. For all $\mu > 0$ and $\lambda \in \Lambda$ the coefficients $c_{41}(\mu,\lambda)$, $c_{42}(\mu,\lambda)$ satisfy

$$(5.7) \qquad \rho^{-1}(\infty)\ q_+\ |c_{42}|^2 = \rho^{-1}(\infty)\ q_+\ |c_{41}|^2 + \rho^{-1}(-\infty)\ q_-.$$

In particular, $c_{42}(\mu,\lambda) \neq 0$.

The proofs of Lemma 5.1 and subsequent lemmas are given at the end of the section.

The asymptotic forms (1.19) and (5.6) coincide if the coefficients satisfy

$$(5.8) \qquad c_+ = a_+\ c_{42},\ c_+\ T_+ = a_+,\ c_+\ R_+ = a_+\ c_{41}.$$

In particular, the first relation and (5.3) imply that

(5.9) $$c_+(\mu,\lambda) = a_+(\mu,\lambda)\,\frac{\rho(\infty)[\phi_4\phi_1](\mu,\lambda)}{2iq_+(\mu,\lambda)}\,.$$

The normalizing factor $a_+(\mu,\lambda)$ will be calculated in §6. The factors $R_+(\mu,\lambda)$, $T_+(\mu,\lambda)$ of (1.19) are independent of the normalization. Indeed, on combining (5.8), (5.9) and (5.3) one finds

(5.10) $$T_+(\mu,\lambda) = \frac{2iq_+(\mu,\lambda)}{\rho(\infty)[\phi_4\phi_1](\mu,\lambda)}\,,$$

(5.11) $$R_+(\mu,\lambda) = -\,\frac{[\phi_4\phi_2](\mu,\lambda)}{[\phi_4\phi_1](\mu,\lambda)}\,.$$

Note that the denominator $[\phi_4\phi_1]$ is not zero by Lemma 5.1 and relations (5.3).

The asymptotic behavior of ψ_- may be discussed similarly. Equations (5.5), (5.1) and Theorem 3.1 imply

(5.12) $$\psi_-(y,\mu,\lambda) = a_-\begin{cases} e^{iq_+y} + o(1), & y \to +\infty,\\ c_{13}\,e^{iq_-y} + c_{14}\,e^{-iq_-y} + o(1), & y \to -\infty,\end{cases}$$

and one has

<u>Lemma 5.2</u>. For all $\mu > 0$ and $\lambda \in \Lambda$ the coefficients $c_{13}(\mu,\lambda)$, $c_{14}(\mu,\lambda)$ satisfy

(5.13) $$\rho^{-1}(-\infty)\,q_-\,|c_{13}|^2 = \rho^{-1}(-\infty)\,q_-\,|c_{14}|^2 + \rho^{-1}(\infty)\,q_+.$$

Comparison of (1.20) and (5.12) gives

(5.14) $$c_- = a_-\,c_{13},\quad c_-\,T_- = a_-,\quad c_-\,R_- = a_-\,c_{14}.$$

Solving these equations for c_-, T_- and R_- and using (5.3) gives

(5.15) $$c_-(\mu,\lambda) = a_-(\mu,\lambda)\,\frac{\rho(-\infty)[\phi_4\phi_1](\mu,\lambda)}{2iq_-(\mu,\lambda)}$$

(5.16) $$T_-(\mu,\lambda) = \frac{2iq_-(\mu,\lambda)}{\rho(-\infty)[\phi_4\phi_1](\mu,\lambda)}\,,$$

(5.17) $$R_-(\mu,\lambda) = -\,\frac{[\phi_1\phi_3](\mu,\lambda)}{[\phi_1\phi_4](\mu,\lambda)}\,.$$

The Spectral Interval Λ_0. For $\lambda \in \Lambda_0$, Theorem 3.1 and the relations (5.1) imply that the only bounded solutions of $A_\mu\phi = \lambda\phi$ are multiples of $\phi_3(y,\mu,\lambda)$. It will be shown that

$$\psi_0(y,\mu,\lambda) = a_0(\mu,\lambda)\,\phi_3(y,\mu,\lambda) \tag{5.18}$$

has the asymptotic form (1.18). Indeed, (5.18), (5.1) and Theorem 3.1 imply that

$$\psi_0(y,\mu,\lambda) = a_0 \begin{cases} c_{31}\, e^{iq_+ y} + c_{32}\, e^{-iq_+ y} + o(1), & y \to +\infty, \\ e^{q'_- y}\,[1 + o(1)], & y \to -\infty. \end{cases} \tag{5.19}$$

The equivalence of (5.19) and (1.18) follows from

Lemma 5.3. For all $\mu > 0$ and $\lambda \in \Lambda_0$ one has

$$c_{32}(\mu,\lambda) = \overline{c_{31}(\mu,\lambda)} \neq 0. \tag{5.20}$$

Comparison of (1.18) and (5.19) gives

$$c_0 = a_0\, c_{32},\quad c_0 T_0 = a_0,\quad c_0\, R_0 = a_0\, c_{31}. \tag{5.21}$$

Solving for c_0, T_0 and R_0 and using (5.3) gives

$$c_0(\mu,\lambda) = a_0(\mu,\lambda)\,\frac{\rho(\infty)\,[\phi_3\phi_1](\mu,\lambda)}{2iq_+(\mu,\lambda)}, \tag{5.22}$$

$$T_0(\mu,\lambda) = \frac{2iq_+(\mu,\lambda)}{\rho(\infty)\,[\phi_3\phi_1](\mu,\lambda)}, \tag{5.23}$$

$$R_0(\mu,\lambda) = -\,\frac{[\phi_3\phi_2](\mu,\lambda)}{[\phi_3\phi_1](\mu,\lambda)}. \tag{5.24}$$

The denominator $[\phi_3\phi_1] \neq 0$ by Lemma 5.3 and relations (5.3).

Finally, note that the conservation laws (1.31) hold; i.e.,

$$\frac{q_\pm}{\rho(\pm\infty)}\,|R_\pm|^2 + \frac{q_\mp}{\rho(\mp\infty)}\,|T_\pm|^2 = \frac{q_\pm}{\rho(\pm\infty)} \text{ for all } \lambda \in \Lambda, \tag{5.25}$$

$$|R_0| = 1 \text{ for all } \lambda \in \Lambda_0. \tag{5.26}$$

In fact, relations (5.25) are equivalent to relations (5.7) and (5.13) of

Lemmas 5.1 and 5.2, as may be seen by combining (5.7), (5.8) and (5.13), (5.14). Similarly, relation (5.26) follows from Lemma 5.3 because (5.21) implies that $R_0 = c_{31}/c_{32}$.

Proof of Lemma 5.1. Relation (5.7) can be verified by calculating $[\phi_4\overline{\phi}_4]$ in two ways, using the asymptotic forms of ϕ_4 as $y \to \infty$ and $y \to -\infty$. Note that for $\lambda \in \Lambda$ one has $\overline{\phi}_4 = \phi_3$ by the uniqueness theorem (Corollary 3.5) and Theorem 3.1. Hence, calculating $[\phi_4\overline{\phi}_4]$ at $y = -\infty$ gives

$$[\phi_4\overline{\phi}_4] = [\phi_4\phi_3] = 2i\,\rho^{-1}(-\infty)q_-, \tag{5.27}$$

by (3.25). Next, relations (5.1) and Theorem 3.1 give (with the notation c.c. for complex conjugate)

$$\begin{aligned} [\phi_4\overline{\phi}_4] &= \phi_4\{\rho^{-1}\overline{\phi}_4'\} - \text{c.c.} \\ &= (c_{41}\phi_1 + c_{42}\phi_2)\{\overline{c}_{41}\,\rho^{-1}\overline{\phi}_1' + \overline{c}_{42}\,\rho^{-1}\overline{\phi}_2'\} - \text{c.c.} \\ &= (c_{41}e^{iq_+y} + c_{42}e^{-iq_+y}) \\ &\quad \times \{\overline{c}_{41}\rho^{-1}(\infty)(-iq_+e^{-iq_+y}) + \overline{c}_{42}\rho^{-1}(\infty)(iq_+e^{iq_+y})\} - \text{c.c.} + o(1) \\ &= \rho^{-1}(\infty)(-iq_+)\{|c_{41}|^2 - c_{41}\overline{c}_{42}e^{2iq_+y} + \overline{c}_{41}c_{42}e^{-2iq_+y} - |c_{42}|^2\} - \text{c.c.} + o(1) \\ &= -2iq_+\rho^{-1}(\infty)\{|c_{41}|^2 - |c_{42}|^2\}. \end{aligned} \tag{5.28}$$

Combining (5.27) and (5.28) gives (5.7).

Proof of Lemma 5.2. (5.13) can be verified by calculating $[\phi_1\overline{\phi}_1]$ in two ways, in analogy with the proof of Lemma 5.1. It can also be derived directly from (5.7) and the relations (5.3).

Proof of Lemma 5.3. Note that for $\lambda \in \Lambda_0$ the uniqueness theorem (Corollary 3.5) and Theorem 3.1 imply that ϕ_3 is real valued and $\overline{\phi}_1 = \phi_2$. Hence relations (5.3) imply

$$\overline{c}_{31} = \rho(\infty)\,\overline{[\phi_3\phi_2]}/2iq_+ = \rho(\infty)[\phi_3\phi_1]/2iq_+ = c_{32}. \tag{5.29}$$

Moreover, if $c_{32} = 0$ then $c_{31} = 0$ by (5.29) and hence by (5.3) one has

(5.30) $$[\phi_3\phi_1] = [\phi_3\phi_2] = 0.$$

But this would imply that ϕ_1 and ϕ_2 are linearly dependent which contradicts (3.25). Hence $c_{32} \neq 0$.

§6. THE SPECTRAL FAMILY OF A_μ

The eigenfunctions and generalized eigenfunctions of §4 and §5 are used in this section to construct the spectral family $\{\Pi_\mu(\lambda)\}$ of A_μ. The construction is based on the Weyl-Kodaira-Titchmarsh theory as presented in Appendix 1. Note that the operator A_μ has the form (A.1) with $I = R$, $p(y) = \rho(y)$, $q(y) = \mu^2 \rho^{-1}(y)$ and $w(y) = c^{-2}(y)\rho^{-1}(y)$. It is clear that p, q and w satisfy (A.2), (A.3), (A.4) when $\rho(y)$ and $c(y)$ are Lebesgue measurable and satisfy (2.1.4).

It will be convenient to decompose R into the disjoint union

(6.1) $$R = \Lambda_d \cup \{c^2(\infty)\mu^2\} \cup \Lambda_0 \cup \{c^2(-\infty)\mu^2\} \cup \Lambda$$

if $c(\infty) < c(-\infty)$ and

(6.2) $$R = \Lambda_d \cup \{c^2(\infty)\mu^2\} \cup \Lambda$$

if $c(\infty) = c(-\infty)$ where

(6.3) $$\Lambda_d = \Lambda_d(\mu) = (-\infty, c^2(\infty)\mu^2),$$

and Λ_0 and Λ are defined by (5.4). The spectral measures of the components of (6.1) and (6.2) will be studied separately.

The Spectral Family in Λ. The spectral measure $\Pi_\mu(\Delta)$ of intervals $\Delta = (a,b) \subset \Lambda$ will be calculated by applying the Weyl-Kodaira theorem to A_μ in Λ. The solution pair

(6.4) $$\begin{cases} \psi_1(y,\lambda) = \phi_4(y,\mu,\lambda) \\ \psi_2(y,\lambda) = \phi_1(y,\mu,\lambda) \end{cases}$$

will be used to obtain a spectral representation in terms of the generalized eigenfunctions $\psi_\pm(y,\mu,\lambda)$ defined by (5.5). The normalizing factors $a_\pm(\mu,\lambda)$ will be chosen after the matrix measure for (ψ_1,ψ_2) has been determined. Note that the pair (ψ_1,ψ_2) satisfies the hypotheses of the Weyl-Kodaira theorem. Indeed, (A.11) follows from Corollary 3.2 and (A.12) from Lemma 5.1.

The Weyl-Kodaira theorem implies that

$$\|\Pi_\mu(\Delta)f\|^2 = \int_\Delta \overline{\hat{f}_j(\lambda)}\, \hat{f}_k(\lambda)\, m_{jk}(d\lambda), \quad \Delta \subset \Lambda, \tag{6.5}$$

for all $f \in \mathcal{H}(R)$ where $(m_{jk}(\Delta))$ is the spectral measure on Λ associated with the basis (6.4) and

$$\hat{f}_j(\lambda) = \lim_{M\to\infty} \int_{-M}^{M} \overline{\psi_j(y,\lambda)}\, f(y)w(y)dy, \tag{6.6}$$

the integrals converging in $L_2(\Lambda,m)$. Thus to complete the determination of $\Pi_\mu(\Delta)$ for $\Delta \subset \Lambda$ it is only necessary to calculate $\{m_{jk}(\Delta)\}$. Now $\Pi_\mu(\Delta)$ can be calculated from the resolvent

$$R_\mu(\zeta) = (A_\mu - \zeta)^{-1} \tag{6.7}$$

by means of Stone's theorem (see, e.g., [25, p. 79]). For $\Delta \subset \Lambda$ the theorem takes the form

$$\|\Pi_\mu(\Delta)f\|^2 = \lim_{\varepsilon\to 0+} \frac{1}{2\pi i} \int_\Delta (f,[R_\mu(\lambda+i\varepsilon) - R_\mu(\lambda-i\varepsilon)]f)d\lambda \tag{6.8}$$

because $\sigma_0(A_\mu) \cap \Lambda = \phi$ by Lemma 4.1. Moreover, $R_\mu(\zeta)$ is an integral operator in $\mathcal{H}(R)$ whose kernel, the Green's function of A_μ, can be represented by the analytic continuation of the basis (6.4) into the ζ-plane. This procedure, whose details are presented in the proofs at the end of the section, leads to

Theorem 6.1. For all $f \in \mathcal{H}(R)$ and all $\Delta \subset \Lambda$ the spectral measure $\Pi_\mu(\Delta)$ satisfies

$$\|\Pi_\mu(\Delta)f\|^2 = \int_\Delta \{A_+^2(\mu,\lambda)\ |\hat{f}_1(\lambda)|^2 + A_-^2(\mu,\lambda)\ |\hat{f}_2(\lambda)|^2\}d\lambda \tag{6.9}$$

where

$$A_\pm^2(\mu,\lambda) = \frac{q_\pm(\mu,\lambda)}{\pi\rho(\pm\infty)\,|[\phi_1\phi_4]|^2}\ . \tag{6.10}$$

Corollary 6.2. The matrix measure $(m_{jk}(\Delta))$ for the basis (6.4) is given by

(6.11) $$m_{11}(\Delta) = \int_\Delta A_+^2(\mu,\lambda)d\lambda, \quad m_{22}(\Delta) = \int_\Delta A_-^2(\mu,\lambda)d\lambda$$

and $m_{12}(\Delta) = m_{21}(\Delta) = 0$ for all $\Delta \subset \Lambda$.

These results suggest an appropriate choice of the normalization factors $a_\pm(\mu,\lambda)$ of (5.5). Note that if instead of the basis (6.4) one takes (5.5) then (6.9) becomes

(6.12) $$\|\Pi_\mu(\Delta)f\|^2 = \int_\Delta \{A_+^2 \; |a_+|^{-2} \; |\hat{f}_+|^2 + A_-^2 \; |a_-|^{-2} \; |\hat{f}_-|^2\}d\lambda$$

where

(6.13) $$\hat{f}_\pm(\mu,\lambda) = \lim_{M\to\infty} \int_{-M}^{M} \overline{\psi_\pm(y,\mu,\lambda)} \; f(y)w(y)dy$$

converge in the space $L_2(\Lambda,m')$ corresponding to (ψ_+,ψ_-). This suggests the choice $|a_\pm|^2 = A_\pm^2$ or

(6.14) $$a_\pm(\mu,\lambda) = e^{i\theta_\pm(\mu,\lambda)} A_\pm(\mu,\lambda)$$

where $\theta_\pm(\mu,\lambda)$ is an arbitrary real valued continuous function. The matrix measure $\{m'_{jk}\}$ for (ψ_+,ψ_-) is independent of the choice of the phase factors $\exp\{i\theta_\pm(\mu,\lambda)\}$ and one could take $\theta_\pm(\mu,\lambda) \equiv 0$. However, it will be more convenient to choose $\theta_\pm(\mu,\lambda)$ in §9 in a way that simplifies the asymptotic form of the normal mode functions $\phi_\pm(y,p,q)$. Theorem 6.1 and the above remarks imply

Corollary 6.3. If the basis $(\psi_+(y,\mu,\lambda),\psi_-(y,\mu,\lambda))$ is normalized by (6.14) then for all $\Delta \subset \Lambda$ one has

(6.15) $$\|\Pi_\mu(\Delta)f\|^2 = \int_\Delta (|\hat{f}_+(\mu,\lambda)|^2 + |\hat{f}_-(\mu,\lambda)|^2)d\lambda$$

and the matrix measure $(m'_{jk}(\Delta))$ for (ψ_+,ψ_-) is given by

(6.16) $$m'_{11}(\Delta) = m'_{22}(\Delta) = |\Delta|$$

and $m'_{12}(\Delta) = m'_{21}(\Delta) = 0$ where $|\Delta|$ is the Lebesgue measure of Δ.

The Spectral Family in Λ_0. The spectral measure of intervals $\Delta \subset \Lambda_0$ will be calculated by applying the Weyl-Kodaira theorem to A_μ, Λ_0 and the solution pair

$$\begin{cases} \psi_1(y,\lambda) = \phi_3(y,\mu,\lambda), \\ \psi_2(y,\lambda) = \phi_1(y,\mu,\lambda). \end{cases} \tag{6.17}$$

The function ϕ_3 is chosen to obtain a representation in terms of the generalized eigenfunction ψ_0 defined by (5.18). The second function could be replaced by any independent solution of $A_\mu\phi = \lambda\phi$. The pair (6.17) satisfies (A.11) by Corollary 3.2 and (A.13) by Lemma 5.3.

Calculation of the spectral measure in Λ_0 by the method described above leads to

Theorem 6.4. For all $f \in \mathcal{H}(R)$ and all $\Delta \subset \Lambda_0$ one has

$$\|\Pi_\mu(\Delta)f\|^2 = \int_\Delta A_0^2(\mu,\lambda)\ |\hat{f}_1(\lambda)|^2\ d\lambda \tag{6.18}$$

where

$$A_0^2(\mu,\lambda) = \frac{q_+(\mu,\lambda)}{\pi\rho(\infty)\,|[\phi_1,\phi_3]|^2}\ . \tag{6.19}$$

Hence the matrix measure $(m_{jk}(\Delta))$ associated with the basis (6.17) is given by

$$m_{11}(\Delta) = \int_\Delta A_0^2(\mu,\lambda)d\lambda \tag{6.20}$$

and $m_{12}(\Delta) = m_{21}(\Delta) = m_{22}(\Delta) = 0$ for all $\Delta \subset \Lambda_0$.

On replacing (6.17) by the basis $(\psi_0(y,\mu,\lambda),\phi_1(y,\mu,\lambda))$ and defining the normalizing factor by

$$a_0(\mu,\lambda) = e^{i\theta_0(\mu,\lambda)}\ A_0(\mu,\lambda) \tag{6.21}$$

where $\theta_0(\mu,\lambda)$ is an arbitrary real valued continuous function one obtains

Corollary 6.5. If $a_0(\mu,\lambda)$ is defined by (6.21) then for all $\Delta \subset \Lambda_0$ one has

$$\|\Pi_\mu(\Delta)f\|^2 = \int_\Delta |\hat{f}_0(\mu,\lambda)|^2\ d\lambda \tag{6.22}$$

where

$$\hat{f}_0(\mu,\lambda) = \lim_{M\to\infty} \int_{-M}^{M} \overline{\psi_0(y,\mu,\lambda)}\ f(y)w(y)dy. \tag{6.23}$$

In particular, the matrix measure $(m'_{jk}(\Delta))$ for the pair (ψ_0, ϕ_1) is given by

$$m'_{11}(\Delta) = |\Delta| \tag{6.24}$$

and $m'_{12}(\Delta) = m'_{21}(\Delta) = m'_{22}(\Delta) = 0$ for all $\Delta \subset \Lambda_0$, and the integral in (6.23) converges in $L_2(\Lambda_0)$.

The Spectral Family in Λ_d. The portion of $\sigma(A_\mu)$ in Λ_d was shown in §4 to be the set of eigenvalues $\{\lambda_k(\mu) \mid 1 \leq k < N(\mu)\}$. Moreover, each $\lambda_k(\mu)$ is a simple eigenvalue with normalized eigenfunction $\psi_k(y,\mu)$ defined by (4.4), and corresponding orthogonal projection $P_{\mu k}$ defined by

$$P_{\mu k}\, f(y) = (\psi_k(\cdot,\mu), f)\, \psi_k(y,\mu). \tag{6.25}$$

Hence, recalling that by convention $\Pi_\mu(\lambda) = \Pi_\mu(\lambda + 0)$, one has

$$\Pi_\mu(\lambda)\, f(y) = \sum_{\lambda_k(\mu) \leq \lambda} (\psi_k(\cdot,\mu), f)\, \psi_k(y,\mu), \quad \lambda \in \Lambda_d. \tag{6.26}$$

The notation in (6.26) denotes a summation over all indices k such that $\lambda_k(\mu) \leq \lambda$. The sum in (6.26) is finite for all $\lambda \in \Lambda_d$.

The Spectral Measure of the Points $c^2(\infty)\mu^2$ and $c^2(-\infty)\mu^2$. When $c(\infty) < c(-\infty)$ one always has $\Pi_\mu(\{c^2(-\infty)\mu^2\}) = 0$ because in this case the special solutions $\phi_1(y,\mu,\lambda)$, $\phi_2(y,\mu,\lambda)$ are defined for $\lambda = c^2(-\infty)\mu^2$ and Theorem 3.1 implies that $A_\mu\phi = c^2(-\infty)\mu^2\phi$ has no solutions in $\mathcal{H}(R)$. The point $\lambda = c^2(\infty)\mu^2$ may be an eigenvalue of A_μ. In each case this question must be decided by determining the behavior of solutions of $A_\mu\phi = c^2(\infty)\mu^2\phi$ for $y \to \pm\infty$. Theorem 3.6 gives simple sufficient conditions for $\Pi_\mu(\{c^2(\infty)\mu^2\}) = 0$. For simplicity, it will be assumed in the remainder of the report that $c^2(\infty)\mu^2 \notin \sigma_0(A_\mu)$. In cases where $c^2(\infty)\mu^2$ is an eigenvalue a corresponding term must be added to the eigenfunction expansion.

The Eigenfunction Expansion for A_μ. Combining the representations of Π_μ obtained above, one finds the representation

$$\begin{aligned} \|\Pi_\mu(\lambda) f\|^2 &= \sum_{k=1}^{N(\mu)-1} H(\lambda - \lambda_k(\mu))\, |\hat{f}_k(\mu)|^2 \\ &+ \int_{\Lambda_0} H(\lambda - \lambda')\, |\hat{f}_0(\mu,\lambda')|^2\, d\lambda' \\ &+ \int_{\Lambda} H(\lambda - \lambda')(|\hat{f}_+(\mu,\lambda')|^2 + |\hat{f}_-(\mu,\lambda')|^2)\, d\lambda' \end{aligned} \tag{6.27}$$

where $H(\lambda) = 1$ for $\lambda \geq 0$, $H(\lambda) = 0$ for $\lambda < 0$,

$$\hat{f}_k(\mu) = \int_R \overline{\psi_k(y,\mu)}\, f(y)\, c^{-2}(y)\, \rho^{-1}(y)dy,\ 1 \leq k < N(\mu), \tag{6.28}$$

$$\hat{f}_0(\mu,\lambda) = L_2(\Lambda_0)\text{-}\lim_{M\to\infty} \int_{-M}^{M} \overline{\psi_0(y,\mu,\lambda)}\, f(y)\, c^{-2}(y)\, \rho^{-1}(y)dy, \tag{6.29}$$

and

$$\hat{f}_\pm(\mu,\lambda) = L_2(\Lambda)\text{-}\lim_{M\to\infty} \int_{-M}^{M} \overline{\psi_\pm(y,\mu,\lambda)}\, f(y)\, c^{-2}(y)\, \rho^{-1}(y)dy. \tag{6.30}$$

In particular, on making $\lambda \to \infty$ one obtains the Parseval relation for A_μ:

$$\begin{aligned} \|f\|^2 = \sum_{k=1}^{N(\mu)-1} |\hat{f}_k(\mu)|^2 + \int_{\Lambda_0} |\hat{f}_0(\mu,\lambda)|^2\, d\lambda\, + \\ + \int_{\Lambda} (|\hat{f}_+(\mu,\lambda)|^2 + |\hat{f}_-(\mu,\lambda)|^2) d\lambda. \end{aligned} \tag{6.31}$$

Thus the correspondence

$$f \to \Psi_\mu f = (\hat{f}_+(\mu,\cdot), \hat{f}_-(\mu,\cdot), \hat{f}_0(\mu,\cdot), \hat{f}_1(\mu), \hat{f}_2(\mu), \cdots) \tag{6.32}$$

defines an isometric mapping Ψ_μ of $\mathcal{H}(R)$ into the direct sum space

$$L_2(\Lambda) + L_2(\Lambda) + L_2(\Lambda_0) + C^{N(\mu)-1} \tag{6.33}$$

and one has

Theorem 6.6. Ψ_μ is a unitary operator from $\mathcal{H}(R)$ to the space (6.33).

This result will be shown to follow from the Weyl-Kodaira theorem and the corresponding properties of the partially isometric operators

$$\begin{aligned} &\Psi_{\mu\pm} : \mathcal{H}(R) \to L_2(\Lambda), \\ &\Psi_{\mu 0} : \mathcal{H}(R) \to L_2(\Lambda_0), \\ &\Psi_{\mu k} : \mathcal{H}(R) \to C,\ 1 \leq k < N(\mu), \end{aligned} \tag{6.34}$$

defined by

$$\Psi_{\mu\pm} f = \hat{f}_\pm(\mu,\cdot),$$

(6.35)
$$\Psi_{\mu 0} f = \hat{f}_0(\mu,\cdot),$$

$$\Psi_{\mu k} f = \hat{f}_k(\mu), \ 1 \le k < N(\mu).$$

In fact, the Weyl-Kodaira theorem implies

Theorem 6.7. The operators (6.34) are partially isometric and if orthogonal projections in $\mathcal{H}(R)$ are defined by

$$P_{\mu\pm} = \Psi^*_{\mu\pm} \Psi_{\mu\pm}, \quad P_{\mu k} = \Psi^*_{\mu k} \Psi_{\mu k}, \ 0 \le k < N(\mu), \tag{6.36}$$

then

$$P_{\mu+} + P_{\mu-} = \Pi_\mu(\Lambda),$$

(6.37)
$$P_{\mu 0} = \Pi_\mu(\Lambda_0),$$

$$\sum_{k=1}^{N(\mu)-1} P_{\mu k} = \Pi_\mu(\Lambda_d).$$

Moreover,

$$\Psi_{\mu\pm} \Psi^*_{\mu\pm} = 1 \text{ in } L_2(\Lambda(\mu)),$$

(6.38)
$$\Psi_{\mu 0} \Psi^*_{\mu 0} = 1 \text{ in } L_2(\Lambda_0(\mu)),$$

$$\Psi_{\mu k} \Psi^*_{\mu k} = 1 \text{ in } C, \ 1 \le k < N(\mu).$$

Corollary 6.8. The inverse isometries $\Psi^*_{\mu\pm}$, $\Psi^*_{\mu k}$, $0 \le k < N(\mu)$, are given by

$$(\Psi^*_{\mu\pm} f_\pm)(y) = \mathcal{H}(R)\text{-}\lim_{\delta\to 0, M\to\infty} \int_{c^2(-\infty)\mu^2+\delta}^{M} \psi_\pm(y,\mu,\lambda)\ f_\pm(\lambda)d\lambda,$$

(6.39)
$$(\Psi^*_{\mu 0} f_0)(y) = \mathcal{H}(R)\text{-}\lim_{\delta\to 0} \int_{c^2(\infty)\mu^2+\delta}^{c^2(-\infty)\mu^2-\delta} \psi_0(y,\mu,\lambda)\ f_0(\lambda)d\lambda,$$

$$(\Psi^*_{\mu k} f_k)(y) = f_k\ \psi_k(y,\mu), \ 1 \le k < N(\mu).$$

The spectral property of the unitary operator Ψ_μ is described by

Corollary 6.9. For every $f \in D(A_\mu)$ one has

$$\Psi_\mu A_\mu f = (\lambda\hat{f}_+(\mu,\lambda),\lambda\hat{f}_-(\mu,\lambda),\lambda\hat{f}_0(\mu,\lambda),\lambda_1(\mu)\hat{f}_1(\mu),\cdots). \tag{6.40}$$

This completes the formulation of the results of §6 and the proofs will now be given.

Proof of Theorem 6.1. The integral representation of the resolvent will be used. Thus, as in the proof of Theorem 4.2,

$$R_\mu(\zeta)\ f(y) = \int_R G_\mu(y,y',\zeta)\ f(y')w(y')dy', \quad \zeta \notin \sigma(A_\mu), \tag{6.41}$$

where $w(y) = c^{-2}(y)\ \rho^{-1}(y)$, and the Green's function G_μ has the form

$$G_\mu(y,y',\zeta) = [\phi_\infty(\cdot,\zeta)\phi_{-\infty}(\cdot,\zeta)]^{-1}\ \phi_{-\infty}(y_<,\zeta)\ \phi_\infty(y_>,\zeta) \tag{6.42}$$

where $y_< = y_<(y,y') = \text{Min}\ (y,y')$, $y_> = y_>(y,y') = \text{Max}\ (y,y')$ and $\phi_{\pm\infty}(y,\zeta)$ are non-trivial solutions of $A_\mu\phi = \zeta\phi$ such that $\phi_{-\infty}(\cdot,\zeta) \in L_2(-\infty,0)$ and $\phi_\infty(\cdot,\zeta) \in L_2(0,\infty)$. To identify these solutions note that by (3.4)

$$\begin{aligned} &\text{Im}\ q_\pm(\mu,\zeta) > 0 \text{ for } \zeta \in R^+(c(-\infty)\mu)^{\text{int}} \\ &\text{Im}\ q_\pm(\mu,\zeta) < 0 \text{ for } \zeta \in R^-(c(-\infty)\mu)^{\text{int}} \end{aligned} \tag{6.43}$$

and

$$e^{\pm y q_\pm'} = e^{\pm i y q_\pm} = e^{y\ \text{Im}\ q_\pm \pm i y\ \text{Re}\ q_\pm}. \tag{6.44}$$

It follows from Theorem 3.1 that

$$\left.\begin{aligned} \phi_\infty(y,\zeta) &= \phi_1(y,\mu,\zeta) \\ \phi_{-\infty}(y,\zeta) &= \phi_4(y,\mu,\zeta) \end{aligned}\right\} \text{ for } \zeta \in R^+(c(-\infty)\mu)^{\text{int}} \tag{6.45}$$

and

$$\left.\begin{aligned} \phi_\infty(y,\zeta) &= \phi_2(y,\mu,\zeta) \\ \phi_{-\infty}(y,\zeta) &= \phi_3(y,\mu,\zeta) \end{aligned}\right\} \text{ for } \zeta \in R^-(c(-\infty)\mu)^{\text{int}}. \tag{6.46}$$

Now the functions appearing in (6.45) and (6.46) have continuous extensions to $R^+(c(-\infty)\mu)$ and $R^-(c(-\infty)\mu)$, respectively, by Corollary 3.2. Indeed, (6.42), (6.45), (6.46) and Corollary 3.2 imply

Lemma 6.10. For all $\lambda \in \Lambda$ one has

$$G_\mu(y,y',\lambda+i0) = [\phi_1\phi_4]^{-1}_{(\lambda)}\ \phi_4(y_<,\mu,\lambda)\ \phi_1(y_>,\mu,\lambda), \tag{6.47}$$

$$G_\mu(y,y',\lambda-i0) = [\phi_2\phi_3]^{-1}_{(\lambda)}\ \phi_3(y_<,\mu,\lambda)\ \phi_2(y_>,\mu,\lambda), \tag{6.48}$$

and the limits are uniform on compact subsets of $R \times R \times \Lambda$.

To prove Theorem 6.1 it is clearly sufficient to verify (6.9) for the functions f of a dense subset of $\mathcal{H}(R)$. It will be convenient to use the subset

$$\mathcal{H}^{com}(R) = \mathcal{H}(R) \cap \{f \mid \text{supp } f \text{ is compact}\}. \tag{6.49}$$

Note that for $f \in \mathcal{H}^{com}(R)$ the integrand in Stone's formula (6.8) can be written

$$\begin{aligned} &\frac{1}{2\pi i}(f,[R_\mu(\lambda+i\varepsilon) - R_\mu(\lambda-i\varepsilon)]f) \\ (6.50)\quad &= \frac{1}{2\pi i}\int_R\int_R \{G_\mu(y,y',\lambda+i\varepsilon) - G_\mu(y,y',\lambda-i\varepsilon)\}\ \overline{f(y)}\ f(y')\ w(y)\ w(y')\ dydy'. \end{aligned}$$

Lemma 6.10 implies that for all $f \in \mathcal{H}^{com}(R)$ this expression tends to a limit

$$(f,H_\mu(\lambda)f) = \int_R\int_R H_\mu(y,y',\lambda)\ \overline{f(y)}\ f(y')\ w(y)\ w(y')\ dydy', \tag{6.51}$$

uniformly on compact subsets $\Delta \subset \Lambda$, where

$$H_\mu(y,y',\lambda) = \frac{1}{2\pi i}\{G_\mu(y,y',\lambda+i0) - G_\mu(y,y',\lambda-i0)\}. \tag{6.52}$$

It follows that

$$\|\Pi_\mu(\Delta)f\|^2 = \int_\Delta (f,H_\mu(\lambda)f)d\lambda \tag{6.53}$$

for all $f \in \mathcal{H}^{com}(R)$. The proof of Theorem 6.1 will be completed by calculating $H_\mu(y,y',\lambda)$. Note that the well-known property $R_\mu(\zeta)^* = R_\mu(\overline{\zeta})$ implies that $\overline{G_\mu(y',y,\zeta)} = G_\mu(y,y',\overline{\zeta})$ and hence

$$G_\mu(y,y',\lambda-i0) = \overline{G_\mu(y',y,\lambda+i0)}. \tag{6.54}$$

Equations (6.52) and (6.54) imply that $H_\mu(y,y',\lambda)$ is Hermitian symmetric:

$$\overline{H_\mu(y',y,\lambda)} = H_\mu(y,y',\lambda). \tag{6.55}$$

Hence, it will be enough to calculate $H_\mu(y,y',\lambda)$ for $y \le y'$.

Calculation of $H_\mu(y,y',\lambda)$. Definition (3.4) of $q_\pm(\mu,\zeta)$ implies that

$$\overline{q_\pm'(\mu,\overline{\zeta})} = -q_\pm'(\mu,\zeta) \text{ for } \zeta \in R(c(-\infty)\mu). \tag{6.56}$$

It follows from Theorem 3.1 and Corollary 3.5 that

$$\left.\begin{aligned} \phi_1(y,\mu,\zeta) &= \overline{\phi_2(y,\mu,\overline{\zeta})} \\ \phi_4(y,\mu,\zeta) &= \overline{\phi_3(y,\mu,\overline{\zeta})} \end{aligned}\right\} \text{ for } \zeta \in R^+(c(-\infty)\mu) \tag{6.57}$$

and

$$\left.\begin{aligned} \phi_2(y,\mu,\zeta) &= \overline{\phi_1(y,\mu,\overline{\zeta})} \\ \phi_3(y,\mu,\zeta) &= \overline{\phi_4(y,\mu,\overline{\zeta})} \end{aligned}\right\} \text{ for } \zeta \in R^-(c(-\infty)\mu). \tag{6.58}$$

It follows that

$$\phi_1(y',\mu,\lambda) = \overline{\phi_2(y',\mu,\lambda)} = \overline{c_{21}(\mu,\lambda)}\ \overline{\phi_1(y',\mu,\lambda)} + \overline{c_{24}(\mu,\lambda)}\ \overline{\phi_4(y',\mu,\lambda)} \tag{6.59}$$

where

$$c_{21} = \frac{[\phi_2\phi_4]}{[\phi_1\phi_4]}, \quad c_{24} = \frac{[\phi_2\phi_1]}{[\phi_4\phi_1]}. \tag{6.60}$$

Similarly

$$\phi_2(y',\mu,\lambda) = \overline{\phi_1(y',\mu,\lambda)} \tag{6.61}$$

and

$$\phi_3(y,\mu,\lambda) = c_{31}(\mu,\lambda)\ \phi_1(y,\mu,\lambda) + c_{34}(\mu,\lambda)\ \phi_4(y,\mu,\lambda) \tag{6.62}$$

where

$$c_{31} = \frac{[\phi_3\phi_4]}{[\phi_1\phi_4]}, \quad c_{34} = \frac{[\phi_3\phi_1]}{[\phi_4\phi_1]}. \tag{6.63}$$

Combining these relations and Lemma 6.10, one finds for $\lambda \in \Lambda$, $y \le y'$

$$\begin{aligned} G_\mu(y,y',\lambda+i0) &= [\phi_1\phi_4]^{-1}\ \phi_4(y,\mu,\lambda)\ \phi_1(y',\mu,\lambda) \\ &= [\phi_1\phi_4]^{-1}\{\bar{c}_{21}\ \phi_4(y,\mu,\lambda)\ \overline{\phi_1(y',\mu,\lambda)} + \bar{c}_{24}\ \phi_4(y,\mu,\lambda)\ \overline{\phi_4(y',\mu,\lambda)}\} \end{aligned} \tag{6.64}$$

and

$$\begin{aligned} G_\mu(y,y',\lambda-i0) &= [\phi_2\phi_3]^{-1}\ \phi_3(y,\mu,\lambda)\ \phi_2(y',\mu,\lambda) \\ &= [\phi_2\phi_3]^{-1}\{c_{31}\ \phi_1(y,\mu,\lambda)\ \overline{\phi_1(y',\mu,\lambda)} + c_{34}\ \phi_4(y,\mu,\lambda)\ \overline{\phi_1(y',\mu,\lambda)}\}. \end{aligned} \tag{6.65}$$

Combining (6.52), (6.64) and (6.65) gives

$$\begin{aligned} H_\mu(y,y',\lambda) = \frac{1}{2\pi i}\Bigg(&\frac{\bar{c}_{24}}{[\phi_1\phi_4]}\ \phi_4(y,\mu,\lambda)\ \overline{\phi_4(y',\mu,\lambda)} \\ &- \frac{c_{31}}{[\phi_2\phi_3]}\ \phi_1(y,\mu,\lambda)\ \overline{\phi_1(y',\mu,\lambda)} \\ &+ \left\{\frac{\bar{c}_{21}}{[\phi_1\phi_4]} - \frac{c_{34}}{[\phi_2\phi_3]}\right\} \phi_4(y,\mu,\lambda)\ \overline{\phi_1(y',\mu,\lambda)}\Bigg) \end{aligned} \tag{6.66}$$

To calculate the coefficients in (6.66) recall that $\overline{\phi_1(y,\mu,\lambda)} = \phi_2(y,\mu,\lambda)$, $\overline{\phi_3(y,\mu,\lambda)} = \phi_4(y,\mu,\lambda)$ for $\lambda \in \Lambda$. It follows from (6.60), (6.63) and (3.25) that

$$\frac{\bar{c}_{24}}{[\phi_1\phi_4]} = \frac{-[\phi_1\phi_2]}{|[\phi_1\phi_4]|^2} = \frac{2i\rho^{-1}(\infty)q_+}{|[\phi_1\phi_4]|^2} = 2\pi i\ A_+^2, \tag{6.67}$$

$$\frac{-c_{31}}{[\phi_2\phi_3]} = \frac{-[\phi_3\phi_4]}{|[\phi_1\phi_4]|^2} = \frac{2i\rho^{-1}(-\infty)q_-}{|[\phi_1\phi_4]|^2} = 2\pi i\ A_-^2, \tag{6.68}$$

$$\frac{\bar{c}_{21}}{[\phi_1\phi_4]} - \frac{c_{34}}{[\phi_2\phi_3]} = \frac{[\phi_1\phi_3]}{[\phi_1\phi_4][\phi_2\phi_3]} - \frac{[\phi_3\phi_1]}{[\phi_4\phi_1][\phi_2\phi_3]} = 0. \tag{6.69}$$

Thus (6.66) can be written

$$
\begin{aligned}
H_\mu(y,y',\lambda) &= A_+^2\, \phi_4(y,\mu,\lambda)\, \overline{\phi_4(y',\mu,\lambda)} + A_-^2\, \phi_1(y,\mu,\lambda)\, \overline{\phi_1(y',\mu,\lambda)} \\
&= A_+^2\, \psi_1(y,\lambda)\, \overline{\psi_1(y',\lambda)} + A_-^2\, \psi_2(y,\lambda)\, \overline{\psi_2(y',\lambda)}.
\end{aligned} \tag{6.70}
$$

This was proved for $y \le y'$. However, both sides of (6.70) are Hermitian symmetric and it therefore holds for all $(y,y') \in R^2$. Multiplying (6.70) by $\overline{f(y)}\, f(y')\, w(y)\, w(y')$ and integrating over R^2 gives

$$
(f, H_\mu(\lambda) f) = A_+^2\, |\hat{f}_1(\lambda)|^2 + A_-^2\, |\hat{f}_2(\lambda)|^2 \tag{6.71}
$$

and combining (6.53) and (6.71) gives (6.9). This completes the proof of Theorem 6.1.

Proof of Corollary 6.2. It follows from (6.5) and (6.9) that

$$
\begin{aligned}
(f, \Pi_\mu(\Delta) g) &= \int_\Delta \overline{\hat{f}_j(\lambda)}\, \hat{g}_j(\lambda)\, m_{jk}(d\lambda) \\
&= \int_\Delta \{A_+^2\, \overline{\hat{f}_1(\lambda)}\, \hat{g}_1(\lambda) + A_-^2\, \overline{\hat{f}_2}(\lambda)\, \hat{g}_2(\lambda)\} d\lambda
\end{aligned} \tag{6.72}
$$

for all $f,g \in \mathcal{H}(R)$. Moreover, the Weyl-Kodaira theorem implies that $\hat{f} = (\hat{f}_1, \hat{f}_2)$ and $\hat{g} = (\hat{g}_1, \hat{g}_2)$ can be arbitrary vectors in $L_2(\Lambda, m)$. Now all vectors $\hat{f}$ with bounded Borel functions as components are in $L_2(\Lambda, m)$. Thus the first of equations (6.11) can be obtained from (6.72) by taking $\hat{f}_1 = \hat{g}_1 = \chi_\Delta$, the characteristic function of Δ, and $\hat{f}_2 = \hat{g}_2 = 0$. The remaining equations of Corollary 6.2 are obtained similarly.

Proof of Theorem 6.4. The proof follows that of Theorem 6.1. Equations (6.41), (6.42) for the resolvent are still valid. However, instead of (6.43) one has

$$
\begin{aligned}
&\operatorname{Im} q_+(\mu,\zeta) > 0 \text{ for } \zeta \in R^+(c(\infty)\mu)^{\text{int}}, \\
&\operatorname{Im} q_+(\mu,\zeta) < 0 \text{ for } \zeta \in R^-(c(\infty)\mu)^{\text{int}}, \\
&\operatorname{Im} q_-(\mu,\zeta) < 0 \text{ for } \zeta \in \mathcal{R}(c(\infty)\mu) \cap \mathcal{L}(c(-\infty)\mu).
\end{aligned} \tag{6.73}
$$

Thus by Theorem 3.1

$$(6.74)\qquad \left.\begin{aligned} \phi_\infty(y,\zeta) &= \phi_1(y,\mu,\zeta) \\ \phi_{-\infty}(y,\zeta) &= \phi_3(y,\mu,\zeta) \end{aligned}\right\} \text{ for } \zeta \in R^+(c(\infty)\mu)^{\text{int}} \cap L(c(-\infty)\mu),$$

and

$$(6.75)\qquad \left.\begin{aligned} \phi_\infty(y,\zeta) &= \phi_2(y,\mu,\zeta) \\ \phi_{-\infty}(y,\zeta) &= \phi_3(y,\mu,\zeta) \end{aligned}\right\} \text{ for } \zeta \in R^-(c(\infty)\mu)^{\text{int}} \cap L(c(-\infty)\mu).$$

Hence Corollary 3.2 implies

Lemma 6.11. For all $\lambda \in \Lambda_0$ and $y \le y'$ one has

$$(6.76)\qquad G_\mu(y,y',\lambda+i0) = [\phi_1\phi_3]^{-1}\, \phi_3(y,\mu,\lambda)\, \phi_1(y',\mu,\lambda),$$

$$(6.77)\qquad G_\mu(y,y',\lambda-i0) = [\phi_2\phi_3]^{-1}\, \phi_3(y,\mu,\lambda)\, \phi_2(y',\mu,\lambda),$$

and the limits are uniform on compact subsets of $R \times R \times \Lambda_0$.

Proceeding to the calculation of $H_\mu(y,y',\lambda)$ one has

$$(6.78)\qquad \left.\begin{aligned} \overline{q_+'(\mu,\overline{\zeta})} &= -q_-'(\mu,\zeta) \\ \overline{q_-'(\mu,\overline{\zeta})} &= q_-'(\mu,\zeta) \end{aligned}\right\} \text{ for } \zeta \in R(c(\infty)\mu) \cap L(c(-\infty)\mu)$$

whence

$$(6.79)\qquad \phi_1(y,\mu,\zeta) = \overline{\phi_2(y,\mu,\overline{\zeta})} \text{ for } \zeta \in R^+(c(\infty)\mu) \cap L(c(-\infty)\mu)$$

and

$$(6.80)\qquad \phi_3(y,\mu,\zeta) = \overline{\phi_3(y,\mu,\overline{\zeta})} \text{ for } \zeta \in R(c(\infty)\mu) \cap L(c(-\infty)\mu).$$

It follows that

$$(6.81)\qquad \phi_1(y',\mu,\lambda) = \overline{\phi_2(y',\mu,\lambda)} = \overline{c_{21}(\mu,\lambda)}\ \overline{\phi_1(y',\mu,\lambda)} + \overline{c_{23}(\mu,\lambda)}\ \overline{\phi_3(y,\mu,\lambda)}$$

where

(6.82) $$c_{21} = \frac{[\phi_2\phi_3]}{[\phi_1\phi_3]} \text{ and } c_{23} = \frac{[\phi_2\phi_1]}{[\phi_3\phi_1]} .$$

These relations and Lemma 6.11 imply that for $\lambda \in \Lambda_0$ and $y \le y'$,

$$G_\mu(y,y',\lambda+i0) = [\phi_1\phi_3]^{-1}\{\bar{c}_{21}\phi_3(y,\mu,\lambda)\overline{\phi_1(y',\mu,\lambda)} + \bar{c}_{23}\phi_3(y,\mu,\lambda)\overline{\phi_3(y',\mu,\lambda)}\}$$
(6.83)

and

(6.84) $$G_\mu(y,y',\lambda-i0) = [\phi_2\phi_3]^{-1}\ \phi_3(y,\mu,\lambda)\ \overline{\phi_1(y',\mu,\lambda)}.$$

Hence

$$H_\mu(y,y',\lambda)$$
(6.85)
$$= \frac{1}{2\pi i}\left\{\frac{\bar{c}_{23}}{[\phi_1\phi_3]}\phi_3(y,\mu,\lambda)\overline{\phi_3(y',\mu,\lambda)} + \left\{\frac{\bar{c}_{21}}{[\phi_1\phi_3]} - \frac{1}{[\phi_2\phi_3]}\right\}\phi_3(y,\mu,\lambda)\overline{\phi_1(y',\mu,\lambda)}\right\}.$$

From the relations $\overline{\phi_1(y,\mu,\lambda)} = \phi_2(y,\mu,\lambda)$, $\overline{\phi_3(y,\mu,\lambda)} = \phi_3(y,\mu,\lambda)$, together with (6.82) and (3.25) it follows that

(6.86) $$\frac{\bar{c}_{23}}{[\phi_1\phi_3]} = \frac{-[\phi_1\phi_2]}{|[\phi_1\phi_3]|^2} = \frac{2i\rho^{-1}(\infty)q_+}{|[\phi_1\phi_3]|^2} = 2\pi i\ A_0^2,$$

and

(6.87) $$\frac{\bar{c}_{21}}{[\phi_1\phi_3]} - \frac{1}{[\phi_2\phi_3]} = \frac{[\phi_1\phi_3]}{[\phi_1\phi_3][\phi_2\phi_3]} - \frac{1}{[\phi_2\phi_3]} = 0.$$

Thus (6.85) can be written

(6.88) $$H_\mu(y,y',\lambda) = A_0^2\ \phi_3(y,\mu,\lambda)\ \overline{\phi_3(y',\mu,\lambda)} = A_0^2\ \psi_1(y,\lambda)\ \overline{\psi_1(y',\lambda)}$$

for $y \le y'$ and hence for all $(y,y') \in R^2$. It follows by integration that for all $f \in \mathcal{H}^{com}(R)$ one has

(6.89) $$(f,H_\mu(\lambda)f) = A_0^2\ |\hat{f}_0(\lambda)|^2,\ \lambda \in \Lambda_0.$$

Combining (6.89) and (6.53) gives (6.18). Finally, (6.18) implies (6.20) by the argument used to prove Corollary 6.2. This completes the proof of Theorem 6.4.

Proof of Theorem 6.6. Ψ_μ is isometric by (6.31), (6.32). Hence to prove that Ψ_μ is unitary it is only necessary to prove that it is surjective. But this is an immediate consequence of Theorem 6.7, equations (6.38). The latter are implied by the Weyl-Kodaira theorem.

Proof of Theorem 6.7 and Corollaries 6.8 and 6.9. These results are direct consequences of the Weyl-Kodaira theorem.

§7. THE DISPERSION RELATIONS FOR THE GUIDED MODES

The eigenvalues of A_μ determine the relation between the wave number $|p|$ and frequency $\omega = \sqrt{\lambda_k(|p|)}$, or dispersion relation, for the guided mode functions $\psi_k(x,y,p)$. The functions $\lambda_k(\mu)$ also appear in the definition (4.4) of $\psi_k(y,\mu)$. The purpose of this section is to provide the information concerning the μ-dependence of $\lambda_k(\mu)$ and $\psi_k(y,\mu)$ that is needed for the spectral analysis of A in §8 and §9.

The domain of definition of the function $\lambda_k(\mu)$ is the set

$$\mathcal{O}_k = \{\mu \mid N(\mu) \geq k + 1\},\ k = 1,2,\cdots. \tag{7.1}$$

Note that $\mathcal{O}_k$ is not empty if and only if $1 \leq k < N_0$ where

$$N_0 = \sup_{\mu>0} N(\mu) \leq +\infty. \tag{7.2}$$

Clearly, if $N_0 < +\infty$ then $N_0 - 1$ is the maximum number of eigenvalues of A_μ for $\mu > 0$. If $N_0 = +\infty$ then either $\sigma_0(A_\mu)$ is infinite for some $\mu > 0$ or $\sigma_0(A_\mu)$ is finite for all $\mu > 0$ and $N(\mu) \to \infty$ when $\mu \to \infty$. Theorem 4.14 implies that both cases occur. The principal result of this section is

Theorem 7.1. For $1 \leq k < N_0$ the set $\mathcal{O}_k$ is open and $\lambda_k : \mathcal{O}_k \to R$ is an analytic function.

The proof of this result given below is based on analytic perturbation theory as developed in [11].

The curves $\lambda = \lambda_k(\mu)$, $\mu \in \mathcal{O}_k$, can never meet or cross because each eigenvalue is simple and the corresponding eigenfunction $\psi_k(y,\mu)$ has exactly k zeros (Theorem 4.8). Thus for $1 \leq k < N_0 - 1$ one has

$$c_m^2\mu^2 \leq \lambda_k(\mu) < \lambda_{k+1}(\mu) < c^2(\infty)\mu^2,\ \mu \in \mathcal{O}_{k+1}. \tag{7.3}$$

Moreover, if $\mathcal{O}_k$ is unbounded then (7.3) implies

$$c_m^2 \leq \liminf_{\mu\to\infty} \mu^{-2} \lambda_k(\mu) \leq \limsup_{\mu\to\infty} \mu^{-2} \lambda_k(\mu) \leq c^2(\infty). \tag{7.4}$$

In particular, if $\mathcal{O}_k$ is unbounded then

$$\lim_{\mu\to\infty} \lambda_k(\mu) = +\infty. \tag{7.5}$$

A related property is given by

Theorem 7.2. For $1 \le k < N_0$ the function $\lambda_k(\mu)$ is strictly monotone increasing; i.e., for all $\mu_1, \mu_2 \in \mathcal{O}_k$ one has

$$\lambda_k(\mu_1) < \lambda_k(\mu_2) \text{ when } \mu_1 < \mu_2. \tag{7.6}$$

The proof of (7.6) given below is based on a variational characterization of $\lambda_k(\mu)$.

By Theorem 7.1, $\mathcal{O}_k$ is open and is therefore a union of disjoint open intervals. Hence the curve $\lambda = \lambda_k(\mu)$ consists of one or more disjoint analytic arcs. It is interesting that these arcs can terminate only on the curve $\lambda = c^2(\infty)\mu^2$. More precisely, one has

Corollary 7.3. Let μ_0 be a boundary point of $\mathcal{O}_k$. Then

$$\lim_{\mu\to\mu_0} \lambda_k(\mu) = c^2(\infty)\mu_0^2. \tag{7.7}$$

It is clear from Theorem 7.2 and (7.3) that the limit in (7.7) exists and does not exceed $c^2(\infty)\mu_0^2$. The equality (7.7) is proved below.

The result (7.4) can be improved by strengthening the hypotheses concerning $c(y)$. A result of this type is

Theorem 7.4. Let $c_m < c(\infty)$ and assume that for each $\varepsilon > 0$ there is an interval $I(\varepsilon) \subset R$ such that

$$c(y) \le c_m + \varepsilon \text{ for all } y \in I(\varepsilon). \tag{7.8}$$

Then $N_0 = +\infty$, $\mathcal{O}_k$ is unbounded for each $k \ge 1$ and $\lambda_k(\mu) \sim c_m^2\mu^2$ when $\mu \to \infty$ in $\mathcal{O}_k$; i.e.,

$$\lim_{\mu\to\infty} \mu^{-2} \lambda_k(\mu) = c_m^2. \tag{7.9}$$

The analyticity of $\lambda_k(\mu)$ and Corollary 3.2 imply the continuity of the eigenfunctions $\psi_k(y,\mu)$. More precisely, one has

Corollary 7.5. For $1 \le k < N_0$ the function $\psi_k(y,\mu)$ satisfies

$$\psi_k,\ \rho^{-1}\psi_k' \in C(R \times \mathcal{O}_k). \tag{7.10}$$

This completes the formulation of the results of §7.

Proof of Theorem 7.1. The analytic perturbation theory of [11, Ch. VII] will be used. Note that the operator A_μ may be defined for all $\mu \in C$ by (1.5), (2.2) and is a closed operator in $\mathcal{H}(R)$. Moreover, the domain $D(A_\mu)$ is independent of μ and is a Hilbert space with respect to the norm defined by

$$\|\phi\|^2_{D(A_\mu)} = \|\phi\|^2_{\mathcal{H}(R)} + \|\phi'\|^2_{\mathcal{H}(R)} + \|(\rho^{-1}\phi')'\|^2_{\mathcal{H}(R)}. \tag{7.11}$$

It follows that $\mu \to A_\mu$ is holomorphic in the generalized sense. Indeed, in the definition of [11, p. 366] one may take $Z = D(A_\mu)$ (independent of μ) and define $U(\mu) : Z \to \mathcal{H}(R)$ to be the identification map. Then $U(\mu)$ is bounded holomorphic (in fact, constant) and

$$V(\mu)\phi = A_\mu U(\mu)\phi = -c^2\rho\ [(\rho^{-1}\phi')' - \rho^{-1}\mu^2\phi] \tag{7.12}$$

is holomorphic for all $\phi \in Z$. Thus A_μ is holomorphic. It follows by [11, p. 370] that each $\lambda_k(\mu)$ has a Puiseux expansion at each point $\mu_0 \in \mathcal{O}_k$. But each $\lambda_k(\mu_0)$ is simple (Lemma 4.3) and hence the Puiseux series can contain no fractional powers of $\mu - \mu_0$ [11, p. 71]. Thus λ_k is in fact analytic at each $\mu_0 \in \mathcal{O}_k$. This proves both statements of Theorem 7.1.

Proof of Theorem 7.2 and Corollary 7.3. The eigenvalue $\lambda_k(\mu)$ can be characterized by the variational principle [5, pp. 1543-4]

$$\lambda_k(\mu) = \inf_{M \in S_k} \sup_{\substack{\phi \in M \cap D(A_\mu) \\ \|\phi\|=1}} (A_\mu\phi,\phi) \tag{7.13}$$

where S_k denotes the set of all k-dimensional subspaces of $\mathcal{H}(R)$. Moreover, $D(A_\mu)$ is independent of μ and

$$(A_\mu\phi,\phi) = \int_R (|\phi'(y)|^2 + \mu^2\ |\phi(y)|^2)\ \rho^{-1}(y)dy \tag{7.14}$$

for all $\phi \in D(A_\mu)$ [11, p. 322]. Hence if $\mu_1 < \mu_2$ then

$$(A_{\mu_1}\phi,\phi) < (A_{\mu_2}\phi,\phi) \tag{7.15}$$

for all $\phi \in D(A_{\mu_1}) = D(A_{\mu_2})$. In particular, (7.13) and (7.15) imply that if $\mu_1, \mu_2 \in \mathcal{O}_k$ then

$$\lambda_k(\mu_1) \le \lambda_k(\mu_2) \tag{7.16}$$

which proves the weak monotonicity of λ_k. It will be convenient to use (7.16) to prove Corollary 7.3 before proving the strong monotonicity.

To prove Corollary 7.3 note that (7.16) and (7.3) imply that the limit in (7.7) exists and does not exceed $c^2(\infty)\mu_0^2$. But if $\lim_{\mu\to\mu_0} \lambda_k(\mu) = \lambda_k^0 < c^2(\infty)\mu_0^2$ then Corollary 3.2 implies that μ_0, λ_k^0 satisfy

(7.17) $$F(\mu_0,\lambda_k^0) = 0;$$

see (4.28). Moreover, $F(\mu,\lambda)$ is analytic at μ_0, λ_k^0 by Corollary 3.3. It follows that λ_k^0 is an eigenvalue of A_{μ_0} and hence $\mu_0 \in \mathcal{O}_k$ by Theorem 7.1. This contradicts the assumption that μ_0 is a boundary point of $\mathcal{O}_k$.

To prove that each $\lambda_k(\mu)$ is strictly monotonic in $\mathcal{O}_k$ two cases will be considered. First, if λ_1, λ_2 are in the same component of $\mathcal{O}_k$, say $(a,b) \subset \mathcal{O}_k$, then $\lambda_k(\mu_1) = \lambda_k(\mu_2)$ would imply that $\lambda_k(\mu) = \text{const.}$ in $[\mu_1,\mu_2]$ and hence in (a,b), since $\lambda_k(\mu)$ by Theorem 7.1. But this contradicts Corollary 7.3 since

(7.18) $$c^2(\infty)a^2 = \lim_{\mu\to a} \lambda_k(\mu) < \lim_{\mu\to b} \lambda_k(\mu) = c^2(\infty)b^2.$$

In the second case μ_1 and μ_2 lie in different components of $\mathcal{O}_k$, say $\mu_1 \in (a_1,b_1) \subset \mathcal{O}_k$ and $\mu_2 \in (a_2,b_2) \subset \mathcal{O}_k$ with $b_1 \le a_2$. In this case, by the preceding argument one has

(7.19) $$\lambda_k(\mu_1) < c^2(\infty)b_1^2 \le c^2(\infty)a_2^2 < \lambda_k(\mu_2)$$

which completes the proof.

Proof of Theorem 7.4 (sketch). The proof is based on the method proposed for the proof of Theorem 4.16 and the variational principle (7.13). Note that the hypothesis (7.8) and Theorem 4.16, generalized to non-constant $\rho(y)$, imply that $N(\mu) \to \infty$ for $\mu \to \infty$. Hence $N_0 = +\infty$ and each $\mathcal{O}_k$ is unbounded.

To prove (7.9) choose piece-wise constant functions $c_0(y)$ and $c^0(y)$ such that

(7.20) $$c_0(y) \le c(y) \le c^0(y) \text{ for all } y \in R,$$

(7.21) $$c_0(y) = c_m < c_0(\infty) \text{ on an interval } I_0, \text{ and}$$

(7.22) $$c_m + \varepsilon = c^0(y) < c^0(\infty) \text{ on an interval } I(\varepsilon).$$

The notation

$$\mathcal{H}_0(R) = L_2(R, c_0^{-2}(y)\ \rho^{-1}(y)dy) \tag{7.23}$$
$$\mathcal{H}^0(R) = L_2(R, c^{0^{-2}}(y)\ \rho^{-1}(y)dy)$$

will be used. The three spaces $\mathcal{H}(R)$, $\mathcal{H}_0(R)$ and $\mathcal{H}^0(R)$ have equivalent norms. In particular, if $\|\phi\|_0$ and $\|\phi\|^0$ denote the norms in $\mathcal{H}_0(R)$ and $\mathcal{H}^0(R)$, respectively, then by (7.20)

$$\|\phi\|^0 \le \|\phi\| \le \|\phi\|_0. \tag{7.24}$$

Now note that the variational principle can be formulated in the homogeneous form

$$\lambda_k(\mu) = \inf_{M\in S_k} \sup_{\substack{\phi\in M\cap D(A_\mu)\\ \phi\neq 0}} \frac{(A_\mu\phi,\phi)}{\|\phi\|^2} \tag{7.25}$$

where

$$\frac{(A_\mu\phi,\phi)}{\|\phi\|^2} = \frac{\int_R (|\phi'|^2 + \mu^2|\phi|^2)\rho^{-1}(y)dy}{\|\phi\|^2}. \tag{7.26}$$

Both the numerator in (7.26) and $D(A_\mu)$ are independent of $c(y)$. Hence, if $A_{0\mu}$ and A_μ^0 denote the operators corresponding to $\rho(y)$, $c_0(y)$ and $\rho(y)$, $c^0(y)$ respectively, then (7.24) and (7.26) imply

$$\frac{(A_{0\mu}\phi,\phi)}{\|\phi\|_0^2} \le \frac{(A_\mu\phi,\phi)}{\|\phi\|^2} \le \frac{(A_\mu^0\phi,\phi)}{\|\phi\|^{0 2}} \tag{7.27}$$

for all $\phi \in D(A_{0\mu}) = D(A_\mu) = D(A_\mu^0)$ such that $\phi \neq 0$. The variational principle (7.25) and (7.27) imply that

$$\lambda_{0k}(\mu) \le \lambda_k(\mu) \le \lambda_k^0(\mu) \tag{7.28}$$

for all sufficiently large μ, where $\{\lambda_{0k}(\mu)\}$ and $\{\lambda_k^0(\mu)\}$ are the eigenvalues of $A_{0\mu}$ and A_μ^0, respectively.

The proof of (7.9) can now be completed by showing by direct calculation that (see [26])

$$\lim_{\mu\to\infty} \mu^{-2}\ \lambda_{0k}(\mu) = c_m^2 \tag{7.29}$$

(7.29 cont.) $$\lim_{\mu\to\infty} \mu^{-2}\,\lambda_k^0(\mu) = (c_m + \varepsilon)^2.$$

It follows from (7.28), (7.29) that

$$c_m^2 \le \liminf_{\mu\to\infty} \mu^{-2}\,\lambda_k(\mu) \le \limsup_{\mu\to\infty} \mu^{-2}\,\lambda_k(\mu) \le (c_m + \varepsilon)^2. \tag{7.30}$$

Equation (7.9) follows because, by hypothesis, $\varepsilon > 0$ is arbitrary.

Proof of Corollary 7.5. The result (7.10) is immediate from Corollary 3.2, the relation

$$\psi_k(y,\mu) = \frac{\phi_3(y,\mu,\lambda_k(\mu))}{\|\phi_3(\cdot,\mu,\lambda_k(\mu))\|} \tag{7.31}$$

and Theorem 7.1.

§8. THE SPECTRAL FAMILY OF A

The acoustic propagator A was defined in Chapter 2 and shown to be a selfadjoint non-negative operator in the Hilbert space $\mathcal{H}$. In this section the spectral family $\{\Pi(\lambda) \mid \lambda \ge 0\}$ of A is constructed by means of the normal mode functions $\{\psi_+, \psi_-, \psi_0, \psi_1, \cdots\}$. The method of construction is to use Fourier analysis in the variables $x \in R^2$ to reduce A to the operator $A_{|p|}$ and then to use the spectral representation of $\Pi_{|p|}(\lambda)$ developed in §6. The construction is given in Theorems 8.1-8.4. In the remainder of the section the proofs of Theorems 8.1-8.4 are developed in a series of lemmas and auxiliary theorems.

The formal definitions of the normal mode functions $\psi_\pm(x,y,p,\lambda)$, $\psi_0(x,y,p,\lambda)$ and $\psi_k(x,y,p)$ were given in §1, equations (1.14)-(1.16) and (1.21)-(1.26). The definitions were completed by the construction of the special solutions $\phi_j(y,\mu,\lambda)$ in §3 and of the normalizing factors $a_\pm(\mu,\lambda)$, $a_0(\mu,\lambda)$, $a_k(\mu)$ in §6. The construction of $\Pi(\lambda)$ will be based on these normal mode functions and the corresponding generalized Fourier transforms. Formally the latter are the scalar products of functions $f \in \mathcal{H}$ with the normal mode functions. The following notation will be used.

$$\tilde{f}_\pm(p,\lambda) = \int_{R^3} \overline{\psi_\pm(x,y,p,\lambda)}\, f(x,y)\, c^{-2}(y)\, \rho^{-1}(y)\, dxdy, \tag{8.1}$$

$$\tilde{f}_0(p,\lambda) = \int_{R^3} \overline{\psi_0(x,y,p,\lambda)}\, f(x,y)\, c^{-2}(y)\, \rho^{-1}(y)\, dxdy, \tag{8.2}$$

$$\tilde{f}_k(p) = \int_{R^3} \overline{\psi_k(x,y,p)}\, f(x,y)\, c^{-2}(y)\, \rho^{-1}(y)\, dxdy,\ k \geq 1. \tag{8.3}$$

Of course these integrals need not converge since the normal mode functions are not in $\mathcal{H}$. Instead, they will be interpreted as Hilbert space limits as in the Plancherel theory of the Fourier transform. This interpretation will be based on the following three theorems.

Theorem 8.1. If $f \in L_1(R^3)$ then the integrals in (8.1), (8.2), (8.3) are absolutely convergent for $(p,\lambda) \in \Omega$, $(p,\lambda) \in \Omega_0$ and $p \in \Omega_k$, respectively, and

$$\tilde{f}_\pm \in C(\Omega),\ \tilde{f}_0 \in C(\Omega_0),\ \tilde{f}_k \in C(\Omega_k),\ k \geq 1. \tag{8.4}$$

For each $f \in \mathcal{H}$ and $M > 0$ define

$$f_M(x,y) = \begin{cases} f(x,y) & \text{if } |x| \leq M \text{ and } |y| \leq M, \\ 0 & \text{if } |x| > M \text{ or } |y| > M. \end{cases} \tag{8.5}$$

It is clear that $f_M \to f$ in $\mathcal{H}$ when $M \to \infty$. Moreover, $f_M \in \mathcal{H} \cap L_1(R^3)$ and one has

Theorem 8.2. For every $f \in \mathcal{H}$ and $M > 0$,

$$\tilde{f}_{M\pm} \in L_2(\Omega),\ \tilde{f}_{M0} \in L_2(\Omega_0),\ \tilde{f}_{Mk} \in L_2(\Omega_k),\ k \geq 1, \tag{8.6}$$

and the Parseval relation holds:

$$\|f_M\|^2_{\mathcal{H}} = \|\tilde{f}_{M+}\|^2_{L_2(\Omega)} + \|\tilde{f}_{M-}\|^2_{L_2(\Omega)} + \sum_{k=0}^{N_0-1} \|\tilde{f}_{Mk}\|^2_{L_2(\Omega_k)}. \tag{8.7}$$

The relation (8.7) suggests the introduction of the direct sum space

$$\tilde{\mathcal{H}} = L_2(\Omega) + L_2(\Omega) + \sum_{k=0}^{N_0-1} L_2(\Omega_k). \tag{8.8}$$

$\tilde{\mathcal{H}}$ is a Hilbert space with norm defined by

$$\|h\|^2_{\tilde{\mathcal{H}}} = \|h_+\|^2_{L_2(\Omega)} + \|h_-\|^2_{L_2(\Omega)} + \sum_{k=0}^{N_0-1} \|h_k\|^2_{L_2(\Omega_k)}; \tag{8.9}$$

see [5, p. 1783]. Theorem 8.2 implies that for each $f \in \mathcal{H}$ and $M > 0$, the sequence $\tilde{f}_M = (\tilde{f}_{M+}, \tilde{f}_{M-}, \tilde{f}_{M0}, \tilde{f}_{M1}, \cdots) \in \tilde{\mathcal{H}}$ and

$$\|f_M\|_{\mathcal{H}} = \|\tilde{f}_M\|_{\tilde{\mathcal{H}}}. \tag{8.10}$$

For arbitrary $f \in \mathcal{H}$ the generalized Fourier transforms associated with A are defined by

Theorem 8.3. For all $f \in \mathcal{H}$, $\{\tilde{f}_M\}$ is a Cauchy sequence in $\tilde{\mathcal{H}}$, for $M \to \infty$, and hence

$$\lim_{M\to\infty} \tilde{f}_M = \tilde{f} = (\tilde{f}_+, \tilde{f}_-, \tilde{f}_0, \tilde{f}_1, \cdots) \tag{8.11}$$

exists in $\tilde{\mathcal{H}}$. In particular, each of the limits

$$\begin{aligned} \tilde{f}_\pm &= L_2(\Omega)\text{-}\lim_{M\to\infty} \tilde{f}_{M\pm} \\ \tilde{f}_0 &= L_2(\Omega_0)\text{-}\lim_{M\to\infty} \tilde{f}_{M0} \\ \tilde{f}_k &= L_2(\Omega_k)\text{-}\lim_{M\to\infty} \tilde{f}_{Mk}, \quad k \geq 1, \end{aligned} \tag{8.12}$$

exists and the Parseval relation

$$\|f\|^2_{\mathcal{H}} = \|\tilde{f}\|^2_{\tilde{\mathcal{H}}} = \|\tilde{f}_+\|^2_{L_2(\Omega)} + \|\tilde{f}_-\|^2_{L_2(\Omega)} + \sum_{k=0}^{N_0-1} \|\tilde{f}_k\|^2_{L_2(\Omega_k)} \tag{8.13}$$

holds for every $f \in \mathcal{H}$.

Theorem 8.3 associates with each $f \in \mathcal{H}$ a family of generalized Fourier transforms $\tilde{f} = (\tilde{f}_+, \tilde{f}_-, \tilde{f}_0, \tilde{f}_1, \cdots) \in \tilde{\mathcal{H}}$ such that

$$\tilde{f}_\pm(p,\lambda) = L_2(\Omega)\text{-}\lim_{M\to\infty} \int_{-M}^{M} \int_{|x|\leq M} \overline{\psi_\pm(x,y,p,\lambda)}\, f(x,y)\, c^{-2}(y)\, \rho^{-1}(y)\, dx dy, \tag{8.14}$$

$$\tilde{f}_0(p,\lambda) = L_2(\Omega_0)\text{-}\lim_{M\to\infty} \int_{-M}^{M} \int_{|x|\leq M} \overline{\psi_0(x,y,p,\lambda)}\, f(x,y) c^{-2}(y)\, \rho^{-1}(y)\, dx dy, \tag{8.15}$$

$$\tilde{f}_k(p) = L_2(\Omega_k)\text{-}\lim_{M\to\infty} \int_{-M}^{M} \int_{|x|\leq M} \overline{\psi_k(x,y,p)}\, f(x,y)\, c^{-2}(y)\, \rho^{-1}(y)\, dx dy, \quad k \geq 1. \tag{8.16}$$

It is easy to verify that if $f \in \mathcal{H} \cap L_1(R^3)$ then the functions $(\tilde{f}_+, \tilde{f}_-, \tilde{f}_0, \tilde{f}_1, \cdots)$ defined by Theorems 8.1 and 8.3 are equivalent and hence the notation is unambiguous. A construction of the spectral family $\{\Pi(\mu)\}$ based on these functions is described by

Theorem 8.4. For all $f, g \in \mathcal{H}$ and all real $\mu \geq 0$, $\Pi(\mu)$ satisfies the relation

$$
(8.17)\qquad
\begin{aligned}
(f,\Pi(\mu)g) &= \int_{\Omega} H(\mu-\lambda)\ \overline{(\tilde f_+(p,\lambda)}\ \tilde g_+(p,\lambda) + \overline{\tilde f_-(p,\lambda)}\ \tilde g_-(p,\lambda)dpd\lambda \\
&+ \int_{\Omega_0} H(\mu-\lambda)\ \overline{\tilde f_0(p,\lambda)}\ \tilde g_0(p,\lambda)dpd\lambda \\
&+ \sum_{k=1}^{N_0-1} \int_{\Omega_k} H(\mu-\lambda_k(|p|))\ \overline{\tilde f_k(p)}\ \tilde g_k(p)dp
\end{aligned}
$$

where $H(\mu) = 1$ for $\mu \geq 0$ and $H(\mu) = 0$ for $\mu < 0$.

The remainder of §8 presents the proofs of these theorems. The proof of Theorem 8.1 will be based on

Lemma 8.5. The normal mode functions satisfy

$$
(8.18)\qquad
\begin{aligned}
&\psi_\pm(x,y,p,\lambda) \in C(R^3 \times \Omega), \\
&\psi_0(x,y,p,\lambda) \in C(R^3 \times \Omega_0), \\
&\psi_k(x,y,p) \in C(R^3 \times \Omega_k),\ k \geq 1.
\end{aligned}
$$

Moreover, for each compact set $K \subset \Omega$ there exists a constant M_K such that

$$
(8.19)\qquad |\psi_\pm(x,y,p,\lambda)| \leq M_K \text{ for all } (x,y) \in R^3 \text{ and } (p,\lambda) \in K.
$$

Similarly, for each compact $K \subset \Omega_0$ there exists a constant M_K such that

$$
(8.20)\qquad |\psi_0(x,y,p,\lambda)| \leq M_K \text{ for all } (x,y) \in R^3 \text{ and } (p,\lambda) \in K,
$$

and for each $k \geq 1$ and compact $K \subset \Omega_k$ there exists a constant M_K such that

$$
(8.21)\qquad |\psi_k(x,y,p)| \leq M_K \text{ for all } (x,y) \in R^3 \text{ and } p \in K.
$$

Proof of Lemma 8.5. To prove (8.18) note that, by (1.16), (1.21)

$$
(8.22)\qquad \psi_+(x,y,p,\lambda) = (2\pi)^{-1}\ e^{ip\cdot x}\ a_+(|p|,\lambda)\ \phi_4(y,|p|,\lambda)
$$

where $a_+(\mu,\lambda)$ is defined by (6.14). The continuity of $\phi_4(y,|p|,\lambda)$ on $R \times \Omega$ follows from Corollary 3.2. The continuity of $a_+(|p|,\lambda)$ on Ω follows from (6.10), (6.14) and the assumed continuity of the phase function $\theta_+(\mu,\lambda)$. Thus the continuity of ψ_+ on $R^3 \times \Omega$ follows from (8.22). The proofs for ψ_-

and ψ_0 are similar and will not be given. The continuity of ψ_k, $k \geq 1$, follows from Corollary 7.5.

To prove (8.19) for ψ_+ note that (8.22) and the continuity of $a_+(|p|,\lambda)$ imply that it is enough to prove the existence of a constant M_K such that

(8.23) $$|\phi_4(y,|p|,\lambda)| \leq M_K \text{ for all } y \in R \text{ and } (p,\lambda) \in K.$$

Now the uniformity of the asymptotic estimates (3.11) on the compact sets Γ_4 of Corollary 3.4 implies that

(8.24) $$\phi_4(y,|p|,\lambda) = \exp\{-i\,y\,q_-(|p|,\lambda)\}[1 + o(1)], \quad y \to -\infty,$$

uniformly for $(p,\lambda) \in K$. Hence, there exists a constant y_K such that

(8.25) $$|\phi_4(y,|p|,\lambda)| \leq 2 \text{ for all } y \leq -y_K \text{ and } (p,\lambda) \in K.$$

Similarly, using the relation

(8.26) $$\phi_4(y,\mu,\lambda) = c_{41}(\mu,\lambda)\,\phi_1(y,\mu,\lambda) + c_{42}(\mu,\lambda)\,\phi_2(y,\mu,\lambda)$$

from (5.1), (5.2), (5.3), the continuity of $c_{41}(\mu,\lambda)$ and $c_{42}(\mu,\lambda)$ and the uniformity of the asymptotic estimates for ϕ_1, ϕ_2 when $y \to +\infty$, one finds that there exist constants y_K', M_K' such that

(8.27) $$|\phi_4(y,|p|,\lambda)| \leq M_K' \text{ for all } y \geq y_K' \text{ and } (p,\lambda) \in K.$$

Finally, the continuity of $\phi_4(y,|p|,\lambda)$ on $R \times \Omega$, which follows from Corollary 3.2, implies the existence of a constant M_K'' such that

(8.28) $$|\phi_4(y,|p|,\lambda)| \leq M_K'' \text{ for } -y_K \leq y \leq y_K' \text{ and } (p,\lambda) \in K.$$

Combining (8.25), (8.27) and (8.28) gives the estimate (8.23) with $M_K = \text{Max}\,(2,M_K',M_K'')$.

The proofs of (8.19) for ψ_- and of (8.20) and (8.21) can be given by the same method. This completes the discussion of Lemma 8.5.

Proof of Theorem 8.1. Consider the function $\tilde{f}_+(p,\lambda)$. The absolute convergence of the integral in (8.1) for each $(p,\lambda) \in \Omega$ follows from (8.19). To prove that $\tilde{f}_+ \in C(\Omega)$ let $(p_0,\lambda_0) \in \Omega$ and let $K \subset \Omega$ be compact and contain (p_0,λ_0) in its interior. Then by Lemma 8.5

(8.29) $$|\overline{\psi_+(x,y,p,\lambda)}\, f(x,y)| \le M_K\, |f(x,y)| \text{ for } (x,y) \in R^3,\ (p,\lambda) \in K.$$

Hence the continuity of $\tilde{f}_+$ at (p_0,λ_0) follows from (8.18) and (8.29) by Lebesgue's dominated convergence theorem. The continuity of $\tilde{f}_-$, $\tilde{f}_0$ and $\tilde{f}_k$ follows by the same argument. This completes the proof of Theorem 8.1.

Relationship of A to A_μ. As a preparation for the proofs of Theorems 8.2, 8.3 and 8.4 the operator A will be related to $A_{|p|}$ by Fourier analysis in the variables $x \in R^2$. To this end note that if $u \in \mathcal{H}$ then Fubini's theorem implies that $u(\cdot,y) \in L_2(R^2)$ for almost every $y \in R$. Thus if $F : L_2(R^2) \to L_2(R^2)$ denotes the Fourier transform in $L_2(R^2)$ then the Plancherel theory implies that

(8.30) $$\hat{u}(p,y) = (Fu)(p,y) = L_2(R^2)\text{-}\lim_{M\to\infty} (2\pi)^{-1} \int_{|x|\le M} e^{-ip\cdot x}\, u(x,y)\,dx$$

exists for almost every $y \in R$ and

(8.31) $$\int_{R^2} |\hat{u}(p,y)|^2\, dp = \int_{R^2} |u(x,y)|^2\, dx \text{ for a.e. } y \in R.$$

Another application of Fubini's theorem gives

Lemma 8.6. $u \in \mathcal{H}$ if and only if $\hat{u} = Fu \in \mathcal{H}$ and the mapping $F : \mathcal{H} \to \mathcal{H}$ is unitary. In particular

(8.32) $$\|\hat{u}\|_{\mathcal{H}} = \|u\|_{\mathcal{H}} \text{ for all } u \in \mathcal{H}.$$

The Fourier transform of the acoustic propagator A will be denoted by $\hat{A}$. Thus

(8.33) $$\hat{A} = F\, A\, F^{-1},\ D(\hat{A}) = F\, D(A).$$

A more detailed characterization of $D(\hat{A})$ is needed to relate $\hat{A}$ to $A_{|p|}$. It will be based on

Lemma 8.7. Let $u \in \mathcal{H}$. Then $D_j u \in \mathcal{H}$ $(j = 1,2)$ if and only if $p_j \hat{u}(p,y) \in \mathcal{H}$ and

(8.34) $$F\, D_j u = p_j\, F\, u,\ j = 1,2.$$

Similarly, $D_y u \in \mathcal{H}$ if and only if $D_y \hat{u} \in \mathcal{H}$ and

(8.35) $$F\, D_y u = D_y\, F\, u.$$

Proof of Lemma 8.7. The distributional derivatives $D_j u$, $D_y u$ may be characterized as temperate distributions on the Schwartz space $S(R^3)$ of rapidly decreasing testing functions [10]. $S(R^3)$ is mapped onto itself by F. The proof of (8.34) is essentially the same as in the standard Plancherel theory. To verify (8.35) note that the distribution-theoretic definition of $D_y u \in \mathcal{H}$ is

$$(8.36) \quad \int_{R^3} D_y u(x,y)\ \phi(x,y)\,dxdy = -\int_{R^3} u(x,y)\ D_y\phi(x,y)\,dxdy \text{ for all } \phi \in S(R^3).$$

Application of Parseval's relation gives

$$(8.37) \quad \int_{R^3} (F\ D_y u)\hat{\phi}\ dpdy = -\int_{R^3} (Fu)(F\ D_y\phi)\,dpdy \text{ for all } \phi \in S(R^3).$$

Now for $\phi \in S(R^3)$ it is easy to verify that

$$(8.38) \quad D_y\hat{\phi} = D_y\,F\,\phi = F\ D_y\phi.$$

Thus (8.37) is equivalent to

$$(8.39) \quad \int_{R^3} (F\ D_y u)\hat{\phi}\ dpdy = -\int_{R^3} (Fu)\ D_y\hat{\phi}\ dpdy \text{ for all } \hat{\phi} \in S(R^3)$$

which in turn is equivalent to (8.35).

Application of Lemma 8.7 to $\hat{A}$ gives

Lemma 8.8. The operator $\hat{A}$ is characterized by the relations

$$(8.40) \quad F\ L_2^1(R^3) = \{\hat{u} \mid p_1\hat{u},\ p_2\hat{u} \text{ and } D_y\hat{u} \text{ are in } \mathcal{H}\},$$

$$(8.41) \quad D(\hat{A}) = F\ L_2^1(R^3) \cap \{\hat{u} \mid D_y\ (\rho^{-1}D_y\hat{u}) - |p|^2\ \hat{u} \in \mathcal{H}\}, \text{ and}$$

$$(8.42) \quad \hat{A}\ \hat{u} = -c^2\ \{\rho D_y(\rho^{-1}D_y\hat{u}) - |p|^2\ \hat{u}\},\ \hat{u} \in D(\hat{A}).$$

Proof of Lemma 8.8. These results follow from application of Lemma 8.7 to the definition of $L_2^1(R^3)$, $D(A)$ and A - equations (2.1.12), (2.2.3) and (2.2.4).

Corollary 8.9. For all $u \in D(A)$ one has

$$(8.43) \quad \hat{u}(p,\cdot) \in D(A_{|p|}) \text{ and}$$

$$(\widehat{Au})(p,\cdot) = A_{|p|}\, \hat{u}(p,\cdot) \tag{8.44}$$

for almost every $p \in R^2$.

Corollary 8.9 is an immediate consequence of Lemma 8.6, Lemma 8.8 and Fubini's theorem.

The Sets $\mathcal{H}^{com}$, $\mathcal{H}'$, $\mathcal{H}''$ and $\mathcal{H}'''$. The following subsets of $\mathcal{H}$ will be used in the proofs of Theorems 8.2, 8.3 and 8.4.

$$\mathcal{H}^{com} = \mathcal{H} \cap \{f \mid \text{supp } f \text{ is compact}\}, \tag{8.45}$$

$$\mathcal{H}' = F^{-1}\mathcal{D}(R^3) = \{f \mid \hat{f} = Ff \in \mathcal{D}(R^3)\}, \tag{8.46}$$

$$\mathcal{H}'' = \{(f(x,y) = f_1(x)f_2(y) \mid f_1 \in F^{-1}\mathcal{D}(R^2), f_2 \in \mathcal{H}(R), \text{supp } f_2 \text{ compact}\}, \tag{8.47}$$

$$\mathcal{H}''' = \text{span } \mathcal{H}'' = \left\{f = \sum_{\alpha=1}^{m} a_\alpha f_\alpha \mid a_\alpha \in C, f_\alpha \in \mathcal{H}''\right\}. \tag{8.48}$$

In (8.46) and (8.47), $\mathcal{D}(R^n)$ denotes the Schwartz space of testing functions with compact support [10]. The sets $\mathcal{H}^{com}$, $\mathcal{H}'$ and $\mathcal{H}'''$ are linear submanifolds of $\mathcal{H}$ which are dense in $\mathcal{H}$. Indeed, it is well known that $\mathcal{D}(R^3)$ is dense in $L_2(R^3)$. This fact implies that $\mathcal{H}^{com}$ is dense in $\mathcal{H}$. The denseness of $\mathcal{H}'$ in $\mathcal{H}$ follows from that of $\mathcal{D}(R^3)$ and the unitarity of F. The denseness of $\mathcal{H}'''$ follows from the fact that $\mathcal{H}$ is the tensor product of $L_2(R^2)$ and $\mathcal{H}(R)$.

It is clear that each of the sets $\mathcal{H}^{com}$, $\mathcal{H}'$ and $\mathcal{H}'''$ is a subset of $\mathcal{H} \cap L_1(R^3)$. Hence, for f in one of these sets, the transforms $\tilde{f}_\pm$, $\tilde{f}_0$ and $\tilde{f}_k$ defined by (8.1)-(8.3) are continuous functions by Theorem 8.1. An alternative characterization is given by

Lemma 8.10. If $f \in \mathcal{H}^{com} \cup \mathcal{H}' \cup \mathcal{H}'''$ then

$$\tilde{f}_\pm(p,\lambda) = \int_R \overline{\psi_\pm(y,|p|,\lambda)}\, \hat{f}(p,y)\, c^{-2}(y)\, \rho^{-1}(y)dy, \tag{8.49}$$

$$\tilde{f}_0(p,\lambda) = \int_R \overline{\psi_0(y,|p|,\lambda)}\, \hat{f}(p,y)\, c^{-2}(y)\, \rho^{-1}(y)dy, \tag{8.50}$$

$$\tilde{f}_k(p) = \int_R \overline{\psi_k(y,|p|)}\, \hat{f}(p,y)\, c^{-2}(y)\, \rho^{-1}(y)dy,\ k \geq 1. \tag{8.51}$$

Proof of Lemma 8.10. Equations (8.49)-(8.51) follow from (8.1)-(8.3) on substituting the definitions (1.21)-(1.23) of the normal mode functions and carrying out the x-integration. These operations are justified for $f \in L_1(R^3)$ by Lemma 8.5 and Fubini's theorem.

Corollary 8.11. If $f \in \mathcal{H}''$ and $f(x,y) = f_1(x)\, f_2(y)$ then

$$\tilde{f}_{\pm}(p,\lambda) = \hat{f}_1(p)\tilde{f}_{2\pm}(|p|,\lambda) = [F\,\Phi_{|p|\pm}f](p,\lambda) = [\Phi_{|p|\pm}Ff](p,\lambda), \tag{8.52}$$

$$\tilde{f}_0(p,\lambda) = \hat{f}_1(p)\tilde{f}_{20}(|p|,\lambda) = [F\,\Phi_{|p|0}f](p,\lambda) = [\Phi_{|p|0}Ff](p,\lambda), \tag{8.53}$$

$$\tilde{f}_k(p) = \hat{f}_1(p)\tilde{f}_{2k}(|p|) = [F\,\Phi_{|p|k}f](p) = [\Phi_{|p|k}Ff](p). \tag{8.54}$$

These results follow immediately from Lemma 8.10 and the results of §8. The notation

$$R(T,\zeta) = (T - \zeta)^{-1} \tag{8.55}$$

will be used for the resolvent of T. The proofs of Theorems 8.2, 8.3 and 8.4 will be based on Stone's theorem relating $R(A,\zeta)$ and the spectral family of A, together with the following three lemmas relating A and $A_{|p|}$.

Lemma 8.12. Let $f \in \mathcal{H}''$ and let $u_\zeta = R(A,\zeta)f$ or, equivalently, $\hat{u}_\zeta = R(\hat{A},\zeta)\hat{f}$. Then

$$\hat{u}_\zeta(p,y) = \int_R G_{|p|}(y,y',\zeta)\, \hat{f}(p,y')\, c^{-2}(y')\, \rho^{-1}(y')dy' \tag{8.56}$$

for almost every $(p,y) \in R^3$ where $G_\mu(y,y',\zeta)$ is the Green's function for A_μ.

Proof of Lemma 8.12. It is enough to verify (8.56) for functions $f \in \mathcal{H}''$. Let $f(x,y) = f_1(x)f_2(y)$ be such a function so that $\hat{f}(p,y) = \hat{f}_1(p)f_2(y)$. Now $\hat{u}_\zeta \in D(\hat{A})$ and hence by Corollary 8.9

$$((\hat{A} - \zeta)\hat{u}_\zeta)(p,y) = ((A_{|p|} - \zeta)\hat{u}_\zeta)(p,y) = \hat{f}_1(p)f_2(y) \tag{8.57}$$

for almost every $(p,y) \in R^3$. It follows that

$$\hat{u}_\zeta(p,y) = [R(A_{|p|},\zeta)\hat{f}_1(p)f_2](y) = \hat{f}_1(p)[R(A_{|p|},\zeta)f_2](y) \tag{8.58}$$

which is equivalent to (8.56) because

$$[R(A_{|p|},\zeta)f_2](y) = \int_R G_{|p|}(y,y',\zeta)\, f_2(y')\, c^{-2}(y')\, \rho^{-1}(y')dy'. \tag{8.59}$$

Lemma 8.13. Let $f(x,y) = f_1(x)f_2(y)$ and $g(x,y) = g_1(x)g_2(y)$ be elements of $\mathcal{H}''$ and let $\zeta = \lambda + i\varepsilon$ with $\lambda \in R$ and $\varepsilon > 0$. Then for all $\mu \in R$ one has

$$\begin{aligned}&\int_{-1}^{\mu} (f,[R(A,\zeta) - R(A,\overline{\zeta})]g)d\lambda \\ &= \int_{R^2} \overline{\hat{f}_1(p)}\, \hat{g}_1(p) \int_{-1}^{\mu} (f_2,[R(A_{|p|},\zeta) - R(A_{|p|},\overline{\zeta})]g_2)d\lambda dp.\end{aligned} \tag{8.60}$$

Proof of Lemma 8.13. The Plancherel theory implies that

$$(f[R(A,\zeta) - R(A,\overline{\zeta})]g) = (\hat{f},[R(\hat{A},\zeta) - R(\hat{A},\overline{\zeta})]\hat{g}). \tag{8.61}$$

Combining this and (8.58) gives, by Fubini's theorem

$$\begin{aligned}&(f,[R(A,\zeta) - R(A,\overline{\zeta})]g) \\ &= \int_{R^2} \overline{\hat{f}_1(p)}\hat{g}_1(p)(f_2,[R(A_{|p|},\zeta) - R(A_{|p|},\overline{\zeta}]g_2)dp.\end{aligned} \tag{8.62}$$

Now a standard estimate for the resolvent of a selfadjoint operator [11, p. 272] implies that

$$|\overline{\hat{f}_1(p)}\hat{g}_1(p)(f_2,[R(A_{|p|},\zeta) - R(A_{|p|},\overline{\zeta})]g_2)| \le \frac{2}{\varepsilon} |\overline{\hat{f}_1(p)}\hat{g}_1(p)|\, \|f_2\|\, \|g_2\| \tag{8.63}$$

Thus integrating (8.62) over $-1 \le \lambda \le \mu$ gives (8.60) by Fubini's theorem.

Lemma 8.14. The spectral family $\{\Pi(\mu)\}$ satisfies relation (8.17) of Theorem 8.4 for all $f,g \in \mathcal{H}''$.

Proof of Lemma 8.14. Stone's theorem in its general form is [25, p. 79]

$$\begin{aligned}&\lim_{\varepsilon\to 0+} \frac{1}{\pi i} \int_a^b (f,[R(A,\lambda+i\varepsilon) - R(A,\lambda-i\varepsilon)]g)d\lambda \\ &= (f,[\Pi(b) + \Pi(b-) - \Pi(a) - \Pi(a-)]g).\end{aligned} \tag{8.64}$$

In the present case $\sigma(A) \subset [0,\infty)$ and (8.64) implies

(8.65) $\frac{1}{2}(f,[\Pi(\mu) + \Pi(\mu-)]f) = \frac{1}{2\pi i} \lim_{\varepsilon\to 0+} \int_{-1}^{\mu} (f,[R(A,\lambda+i\varepsilon) - R(A,\lambda-i\varepsilon)]f)d\lambda.$

Moreover, since $\Pi(\mu-) = \lim_{\delta\to 0+} \Pi(\mu - \delta)$ and $\lim_{\delta\to 0+} \Pi((\mu-\delta)-) = \Pi(\mu-)$, (8.65) implies

$$(f,\Pi(\mu-)f) = \frac{1}{2\pi i} \lim_{\delta\to 0+} \lim_{\varepsilon\to 0+} \int_{-1}^{\mu} (f,[R(A,\lambda+i\varepsilon) - R(A,\lambda-i\varepsilon)]f)d\lambda. \tag{8.66}$$

If $f(x,y) = f_1(x)f_2(y) \in \mathcal{H}'$ then combining (8.60) with $g = f$ and (8.66) gives

$$\begin{aligned}&(f,\Pi(\mu-)f)\\ &= \frac{1}{2\pi i} \lim_{\delta\to 0+} \lim_{\varepsilon\to 0+} \int_{R^2} |\hat{f}_1(p)|^2 \int_{-1}^{\mu-\delta} (f_2,[R(A_{|p|},\lambda+i\varepsilon) - R(A_{|p|},\lambda-i\varepsilon)]f_2)d\lambda dp.\end{aligned} \tag{8.67}$$

Now application of the spectral theorem to $A_{|p|}$ gives

$$(f_2,[R(A_{|p|},\lambda+i\varepsilon) - R(A_{|p|},\lambda-i\varepsilon)]f_2) = \int_R \frac{2i\varepsilon}{(\lambda-\lambda')^2+\varepsilon^2} (f_2,\Pi_{|p|}(d\lambda')f_2). \tag{8.68}$$

It follows that

$$\begin{aligned}&\left|\int_{-1}^{\mu-\delta} (f_2,[R(A_{|p|},\lambda+i\varepsilon) - R(A_{|p|},\lambda-i\varepsilon)]f_2)d\lambda\right|\\ &\qquad \le \int_R \left(\int_{-1}^{\mu-\delta} \frac{2\varepsilon}{(\lambda-\lambda')^2+\varepsilon^2} d\lambda\right) (f_2,\Pi_{|p|}(d\lambda')f_2).\end{aligned} \tag{8.69}$$

Moreover,

$$\int_{-1}^{\mu-\delta} \frac{2\varepsilon}{(\lambda-\lambda')^2+\varepsilon^2} d\lambda \le \int_R \frac{2\varepsilon}{(\lambda-\lambda')^2+\varepsilon^2} d\lambda = 2\pi \tag{8.70}$$

for all $\lambda' \in R$ and all $\varepsilon > 0$. Combining (8.69) and (8.70) gives

$$\left|\int_{-1}^{\mu-\delta} (f_2,[R(A_{|p|},\lambda+i\varepsilon) - R(A_{|p|},\lambda-i\varepsilon)]f_2)d\lambda\right| \le 2\pi \|f_2\| \tag{8.71}$$

for all $p \in R^2$ and all $\mu \ge 0$, $\delta > 0$ and $\varepsilon > 0$. In addition Stone's theorem applied to $A_{|p|}$ gives

$$\frac{1}{2\pi i} \lim_{\delta\to 0+} \lim_{\varepsilon\to 0+} \int_{-1}^{\mu-\delta} (f_2,R(A_{|p|},\lambda+i\varepsilon) - R(A_{|p|},\lambda-i\varepsilon)]f_2)d\lambda = (f_2,\Pi_{|p|}(\mu-)f_2). \tag{8.72}$$

Equations (8.67), (8.72) and the estimate (8.71) imply, by Lebesgue's dominated convergence theorem,

$$(8.73) \qquad (f,\Pi(\mu-)f) = \int_{R^2} |\hat{f}_1(p)|^2 \, (f_2,\Pi_{|p|}(\mu-)f_2)dp.$$

It follows by polarization that

$$(8.74) \qquad (f,\Pi(\mu-)g) = \int_{R^2} \overline{\hat{f}_1(p)} \, \hat{g}_1(p)(f_2,\Pi_{|p|}(\mu-)g_2)dp$$

for all $f,g \in \mathcal{H}'$. The same argument applied to (8.65) gives the relation

$$(8.75) \quad (f,[\Pi(\mu) + \Pi(\mu-)]g) = \int_{R^2} \overline{\hat{f}_1(p)} \, \hat{g}_1(p)(f_2,[\Pi_{|p|}(\mu) + \Pi_{|p|}(\mu-)]g_2)dp.$$

Subtracting (8.74) from (8.75) gives

$$(8.76) \qquad (f,\Pi(\mu)g) = \int_{R^2} \overline{\hat{f}_1(p)} \, \hat{g}_1(p)(f_2,\Pi_{|p|}(\mu)g_2)dp$$

for all $f,g \in \mathcal{H}'$. To prove the relation (8.17) for $f,g \in \mathcal{H}'$ note that the construction of Π_μ given by (6.27) implies, by polarization

$$(8.77) \qquad \begin{aligned} (f_2,\Pi_{|p|}(\mu)g_2) &= \int_{\Lambda(|p|)} H(\mu-\lambda)[\overline{\hat{f}_{2+}(|p|,\lambda)} \, \hat{g}_{2+}(|p|,\lambda) + \overline{\hat{f}_{2-}(|p|,\lambda)} \, \hat{g}_{2-}(|p|,\lambda)]d\lambda \\ &+ \int_{\Lambda_0(|p|)} H(\mu-\lambda) \, \overline{\hat{f}_{20}(|p|,\lambda)} \, \hat{g}_{20}(|p|,\lambda)d\lambda \\ &+ \sum_{k=1}^{N(|p|)-1} H(\mu-\lambda_k(|p|)) \, \overline{\hat{f}_{2k}(|p|)} \, g_{2k}(|p|) \end{aligned}$$

for all $p \in R^2$ such that $|p| > 0$, where

$$(8.78) \qquad \begin{aligned} \Lambda(|p|) &= \{\lambda \mid c^2(-\infty)|p|^2 < \lambda\}, \text{ and} \\ \Lambda_0(|p|) &= \{\lambda \mid c^2(\infty)|p|^2 < \lambda < c^2(-\infty)|p|^2\}. \end{aligned}$$

Substituting this into (8.76) and recalling the definitions of Ω, Ω_0 and Ω_k $(k \geq 1)$ gives (8.19) for $f,g \in \mathcal{H}'$. The relation extends immediately to all $f,g \in \mathcal{H}''$ by linearity. This completes the proof of Lemma 8.14.

Proof of Theorem 8.2. Let $f \in \mathcal{H}$ and $M > 0$ be given. Then since $\mathcal{H}''$ is dense in $\mathcal{H}$ there exists a sequence $\{g_n\}$ in $\mathcal{H}''$ such that $g_n \to f_M$ in $\mathcal{H}$. Note that since

$$\|f_M - g_n\|_{\mathcal{H}}^2 = \int_{-M}^{M} \int_{R^2} |f(x,y) - g_n(x,y)|^2 c^{-2}(y) \rho^{-1}(y)dxdy \tag{8.79}$$
$$+ \int_{|y| \geq M} \int_{R^2} |g_n(x,y)|^2 c^{-2}(y) \rho^{-1}(y)dxdy$$

it may be assumed that $g_n(x,y) = 0$ for $|y| > M$. Now Lemma 8.14 implies that Parseval's relation

$$\|g\|_{\mathcal{H}}^2 = \|\tilde{g}\|_{\tilde{\mathcal{H}}}^2 = \|\tilde{g}_+\|^2 + \|\tilde{g}_-\|^2 + \sum_{k=0}^{N_0-1} \|\tilde{g}_k\|^2 \tag{8.80}$$

holds for all $g \in \mathcal{H}''$, where $\tilde{g} = (\tilde{g}_+, \tilde{g}_-, \tilde{g}_0, \tilde{g}_1, \cdots)$. On applying (8.80) to the differences $g_n - g_m$ it is found that $\tilde{g}_n = (\tilde{g}_{n+}, \tilde{g}_{n-}, \tilde{g}_{n0}, \tilde{g}_{n1}, \cdots)$ is a Cauchy sequence in $\tilde{\mathcal{H}}$. Hence there exists a limit

$$\lim_{n\to\infty} \tilde{g}_n = h = (h_+, h_-, h_0, h_1, \cdots) \in \tilde{\mathcal{H}}, \tag{8.81}$$

since $\tilde{\mathcal{H}}$ is a Hilbert space. To complete the proof of Theorem 8.2 it will be enough to show that

$$\begin{aligned} \tilde{f}_{M\pm}(p,\lambda) &= h_{\pm}(p,\lambda) \text{ for a.e. } (p,\lambda) \in \Omega, \\ \tilde{f}_{M0}(p,\lambda) &= h_0(p,\lambda) \text{ for a.e. } (p,\lambda) \in \Omega_0, \\ \tilde{f}_{Mk}(p) &= h_k(p) \text{ for a.e. } p \in \Omega_k,\ k \geq 1. \end{aligned} \tag{8.82}$$

This clearly implies (8.6) since $h \in \tilde{\mathcal{H}}$. Moreover, since Hilbert space convergence implies convergence of the norms, the relation (8.80) for $g_n \in \mathcal{H}''$ implies

$$\|f_M\|_{\mathcal{H}}^2 = \lim_{n\to\infty} \|g_n\|_{\mathcal{H}}^2 = \lim_{n\to\infty} \|\tilde{g}_n\|_{\tilde{\mathcal{H}}}^2 = \|h\|_{\tilde{\mathcal{H}}}^2 = \|h_+\|^2 + \|h_-\|^2 + \sum_{k=0}^{N_0-1} \|h_k\|^2 \tag{8.83}$$

which is equivalent to (8.7) when (8.82) holds.

Relation (8.82) will be proved for $\tilde{f}_{M+}$. The proofs for the remaining cases are entirely similar. To prove that $\tilde{f}_{M+}(p,\lambda) = h_+(p,\lambda)$ for a.e. (p,λ) in Ω note that if K is any compact subset of Ω then $\tilde{f}_{M+} \in C(K) \subset L_2(K)$ by Theorem 8.1 and

$$\|\tilde{f}_{M+} - h_+\|_{L_2(K)} = \lim_{n\to\infty} \|\tilde{f}_{M+} - \tilde{g}_{n+}\|_{L_2(K)}. \tag{8.84}$$

Now by Lemma 8.10

$$\tilde{f}_{M+}(p,\lambda) - \tilde{g}_{n+}(p,\lambda) = \int_{-M}^{M} \overline{\psi_+(y,|p|,\lambda)}\, [\hat{f}_M(p,y) - \hat{g}_n(p,y)]c^{-2}(y)\rho^{-1}(y)dy. \tag{8.85}$$

Hence by Lemma 8.5 and Schwarz's inequality

$$\begin{aligned} |\tilde{f}_{M+}(p,\lambda) - \tilde{g}_{n+}(p,\lambda)| &\le M_K \int_{-M}^{M} |\hat{f}_M(p,y) - \hat{g}_n(p,y)|\; c^{-2}(y)\rho^{-1}(y)dy \\ &\le M_K\left(\int_{-M}^{M} c^{-2}(y)\rho^{-1}(y)dy\right)^{1/2}\left(\int_{-M}^{M} |\hat{f}_M(p,y) - \hat{g}_n(p,y)|^2 c^{-2}(y)\rho^{-1}(y)dy\right)^{1/2} \end{aligned} \tag{8.86}$$

for all $(p,\lambda) \in K$. It follows that there is a constant $C = C(K,M)$ such that

$$\|\tilde{f}_{M+} - \tilde{g}_{n+}\|_{L_2(K)} \le C\, \|\hat{f}_M - \hat{g}_n\|_{\mathcal{H}} = C\, \|f_M - g_n\|_{\mathcal{H}}. \tag{8.87}$$

Since $g_n \to f_M$ in $\mathcal{H}$, (8.87) implies that the limit in (8.84) is zero and hence $\tilde{f}_{M+}(p,\lambda) = h_+(p,\lambda)$ for a.e. $(p,\lambda) \in K$. This completes the proof since $K \subset \Omega$ was an arbitrary compact set.

<u>Proof of Theorem 8.3</u>. To prove that $\{\tilde{f}_M\}$ is a Cauchy sequence in $\tilde{\mathcal{H}}$ for $M \to \infty$, let $M > 0$ and $N > 0$ be arbitrary numbers and let $\{g_n^M\}$, $\{g_n^N\}$ be sequences in $\mathcal{H}''$ such that $g_n^M \to f_M$, $g_n^N \to f_N$ in $\mathcal{H}$. Then, as proved above, $\tilde{g}_n^M \to \tilde{f}_M$ and $\tilde{g}_n^N \to \tilde{f}_N$ in $\tilde{\mathcal{H}}$ and Parseval's relation (8.80) holds for $\tilde{g}_n^M - \tilde{g}_n^N$. Passage to the limit $n \to \infty$ gives

$$\|f_M - f_N\|_{\mathcal{H}}^2 = \|\tilde{f}_M - \tilde{f}_N\|_{\tilde{\mathcal{H}}}^2 \tag{8.88}$$

which implies that $\{\tilde{f}_M\}$ is a Cauchy sequence because $f_M \to f$ in $\mathcal{H}$ when $M \to \infty$. Finally, one gets (8.13) for arbitrary $f \in \mathcal{H}$ by passage to the limit in (8.9).

<u>Proof of Theorem 8.4</u>. It will be enough to prove (8.17) for $f = g \in \mathcal{H}$ since the general case then follows by polarization. Now by Lemma 8.14

$$\begin{aligned} (g,\Pi(\mu)g) &= \int_{\Omega} H(\mu-\lambda)(|\tilde{g}_+(p,\lambda)|^2 + |\tilde{g}_-(p,\lambda)|^2)dpd\lambda \\ &+ \int_{\Omega_0} H(\mu-\lambda)|\tilde{g}_0(p,\lambda)|^2\, dpd\lambda \\ &+ \sum_{k=1}^{N_0-1} \int_{\Omega_k} H(\mu-\lambda_k(|p|))\, |\tilde{g}_k(p)|^2\, dp \end{aligned} \tag{8.89}$$

for all $g \in \mathcal{H}''$. Let $f \in \mathcal{H}$, $M > 0$ and let $\{g_n\}$ be a sequence in $\mathcal{H}''$ such that $g_n \to f_M$ in $\mathcal{H}$. Then it follows from the proof of Theorem 8.2 that $\tilde{g}_n \to \tilde{f}_M$ in $\tilde{\mathcal{H}}$. Replacing g by g_n in (8.89) and making $n \to \infty$ gives (8.89) with $g = f_M$. If $N_0 = +\infty$ then passage to the limit is justified because the right-hand side of (8.89) is majorized by

$$\|\tilde{g}_+\|^2 + \|\tilde{g}_-\|^2 + \sum_{k=0}^{N_0-1} \|\tilde{g}_k\|^2 < \infty. \tag{8.90}$$

Thus (8.89) is valid with $g = f_M$ where $f \in \mathcal{H}$ and $M > 0$ are arbitrary. Making $M \to \infty$ and repeating the above argument gives (8.89) with $g = f \in \mathcal{H}$, by Theorem 8.3.

Another proof may be obtained by noting that the left-hand side of (8.89) is a bounded quadratic form on $\mathcal{H}$, while the right-hand side is a bounded quadratic function of $\tilde{g} = \Psi g$ because of the majorization by (8.90). Thus (8.17) follows from the boundedness of Ψ and the fact that (8.89) holds for g in the dense set $\mathcal{H}''$.

§9. NORMAL MODE EXPANSIONS FOR A

The normal mode expansions for the acoustic propagator A that are the main results of this chapter are formulated and proved in this section. The starting point is the representation of the spectral family of A given by Theorem 8.4. The main result, Theorem 9.8, shows that the family $\{\psi_+, \psi_-, \psi_0, \psi_1, \cdots\}$ is a complete orthogonal family of normal modes for A. Theorem 9.9 shows that it provides a spectral representation of A. These results are shown to imply that the families $\{\phi_+, \psi_1, \psi_2, \cdots\}$ and $\{\phi_-, \psi_1, \psi_2, \cdots\}$, defined in §1, are also complete orthogonal families of normal modes for A and provide alternative spectral representations.

The basic representation space for A associated with the family $\{\psi_+, \psi_-, \psi_0, \psi_1, \cdots\}$ is the direct sum space

$$\tilde{\mathcal{H}} = L_2(\Omega) + L_2(\Omega) + \sum_{k=0}^{N_0-1} L_2(\Omega_k) \tag{9.1}$$

introduced in §8. Theorem 8.3 associates with each $f \in \mathcal{H}$ an element $\tilde{f} \in \tilde{\mathcal{H}}$. The Parseval relation (8.13) implies that the linear operator

$$\Psi : \mathcal{H} \to \tilde{\mathcal{H}} \tag{9.2}$$

defined by

$$\Psi f = \tilde{f} \text{ for all } f \in \mathcal{H} \tag{9.3}$$

is an isometry; i.e.,

$$\|\Psi f\|_{\tilde{\mathcal{H}}} = \|f\|_{\mathcal{H}} \text{ for all } f \in \mathcal{H}. \tag{9.4}$$

The principal result of this section is

Theorem 9.1. The operator Ψ is unitary; i.e.,

$$\Psi^* \Psi = 1 \text{ in } \mathcal{H}, \text{ and} \tag{9.5}$$

$$\Psi \Psi^* = 1 \text{ in } \tilde{\mathcal{H}}. \tag{9.6}$$

Relations (9.5) and (9.6) generalize the completeness and orthogonality properties, respectively, of the eigenfunction expansions for operators with discrete spectra. Relation (9.5) is equivalent to (9.4) and thus follows from Theorem 8.3. Relation (9.6) is shown below to follow from the unitarity of the operator Ψ_μ associated with A_μ (Theorem 6.6).

The completeness relation (9.5) implies that every $f \in \mathcal{H}$ has a normal mode expansion based on the family $\{\psi_+, \psi_-, \psi_0, \psi_1, \cdots\}$. The orthogonality relation (9.6) implies that the space $\tilde{\mathcal{H}}$ is isomorphic to $\mathcal{H}$ and thus provides a parameterization of the set of all states $f \in \mathcal{H}$ of the acoustic field. These implications of Theorem 9.1 will be developed in a series of corollaries.

The normal mode expansion will be based on the linear operators

$$\Psi_\pm : \mathcal{H} \to L_2(\Omega) \tag{9.7}$$

$$\Psi_k : \mathcal{H} \to L_2(\Omega_k), \; 0 \le k < N_0, \tag{9.8}$$

defined for all $f \in \mathcal{H}$ by

$$\Psi_\pm f = \tilde{f}_\pm, \tag{9.9}$$

$$\Psi_k f = \tilde{f}_k, \; 0 \le k < N_0, \tag{9.10}$$

where $\tilde{f}_\pm$, $\tilde{f}_k$ are defined as in Theorem 8.3. It is clear that

$$\Psi f = (\Psi_+ f, \Psi_- f, \Psi_0 f, \Psi_1 f, \cdots) \tag{9.11}$$

for all $f \in \mathcal{H}$ and, by (9.4),

$$\|\Psi_+ f\|^2 + \|\Psi_- f\|^2 + \sum_{k=0}^{N_0-1} \|\Psi_k f\|^2 = \|f\|^2. \tag{9.12}$$

In particular, each of the operators $\Psi_\pm$, Ψ_k is bounded with norm not exceeding 1. The normal mode expansion for A, in abstract form, is given by

Corollary 9.2. The family $\{\Psi_+,\Psi_-,\Psi_0,\Psi_1,\cdots\}$ satisfies

$$1 = \Psi_+^* \Psi_+ + \Psi_-^* \Psi_- + \sum_{k=0}^{N_0-1} \Psi_k^* \Psi_k \tag{9.13}$$

where 1 is the identity operator in $\mathcal{H}$ and the series in (9.13) converges strongly.

It will be shown that (9.13) is equivalent to the completeness relation (9.5). The orthogonality relation (9.6) will be shown to be equivalent to the relations described by

Corollary 9.3. The family $\{\Psi_+,\Psi_-,\Psi_0,\Psi_1,\cdots\}$ satisfies the relations

$$\Psi_\pm \Psi_\pm^* = 1 \text{ in } L_2(\Omega), \tag{9.14}$$

$$\Psi_k \Psi_k^* = 1 \text{ in } L_2(\Omega_k),\ 0 \le k < N_0. \tag{9.15}$$

In addition, $\Psi_+ \Psi_-^* = \Psi_- \Psi_+^* = 0$, $\Psi_\pm \Psi_k^* = 0$, $\Psi_k \Psi_\pm^* = 0$ and $\Psi_k \Psi_\ell^* = 0$ for all k and $\ell \ne k$ such that $0 \le k, \ell < N_0$.

Relations (9.14), (9.15) imply that each of the operators Ψ_+, Ψ_-, Ψ_k $(0 \le k < N_0)$ is partially isometric [11, p. 258]. It follows that the operators in $\mathcal{H}$ defined by

$$\begin{cases} P_\pm = \Psi_\pm^* \Psi_\pm \\ P_k = \Psi_k^* \Psi_k,\ 0 \le k < N_0, \end{cases} \tag{9.16}$$

are orthogonal projections in $\mathcal{H}$ onto subspaces

$$\begin{cases} \mathcal{H}_\pm = P_\pm \mathcal{H} \\ \mathcal{H}_k = P_k \mathcal{H},\ 0 \le k < N_0. \end{cases} \tag{9.17}$$

Combining this with Corollaries 9.2 and 9.3 gives

Corollary 9.4. $\{P_+, P_-, P_0, P_1, \cdots\}$ is a complete family of orthogonal projections in $\mathcal{H}$; i.e., the range spaces defined by (9.17) are mutually orthogonal and

$$1 = P_+ + P_- + \sum_{k=0}^{N_0-1} P_k. \tag{9.18}$$

The spaces (9.17) are subspaces of $\mathcal{H}$ and hence the direct sum space

$$\mathcal{H}_+ + \mathcal{H}_- + \sum_{k=0}^{N_0-1} \mathcal{H}_k \tag{9.19}$$

may be identified with the set of all

$$f = f_+ + f_- + \sum_{k=0}^{N_0-1} f_k \tag{9.20}$$

in $\mathcal{H}$ such that

$$f_\pm \in \mathcal{H}_\pm, \ f_k \in \mathcal{H}_k \text{ for } 0 \le k < N_0 \tag{9.21}$$

and

$$\|f_+\|^2 + \|f_-\|^2 + \sum_{k=0}^{N_0-1} \|f_k\|^2 < \infty. \tag{9.22}$$

With this convention, Corollary 9.4 implies

Corollary 9.5. $\mathcal{H}$ has the decomposition

$$\mathcal{H} = \mathcal{H}_+ + \mathcal{H}_- + \sum_{k=0}^{N_0-1} \mathcal{H}_k. \tag{9.23}$$

The definitions of the operators $\Psi_\pm$, Ψ_k and equations (8.14)-(8.16) imply

Corollary 9.6. The operators $\Psi_\pm$, Ψ_k have the representations

$$(\Psi_\pm f)(p,\lambda) = L_2(\Omega)\text{-}\lim_{M\to\infty} \int_{R_M^3} \overline{\psi_\pm(x,y,p,\lambda)}\, f(x,y) c^{-2}(y) \rho^{-1}(y)\, dxdy, \tag{9.24}$$

$$(\Psi_0 f)(p,\lambda) = L_2(\Omega_0)\text{-}\lim_{M\to\infty} \int_{R_M^3} \overline{\psi_0(x,y,p,\lambda)}\, f(x,y) c^{-2}(y) \rho^{-1}(y)\, dxdy, \tag{9.25}$$

$$(\Psi_k f)(p) = L_2(\Omega_k)\text{-}\lim_{M\to\infty} \int_{R_M^3} \overline{\psi_k(x,y,p)}\, f(x,y) c^{-2}(y) \rho^{-1}(y)\, dxdy, \tag{9.26}$$

for $1 \le k < N_0$ where $R_M^3 = R^3 \cap \{(x,y) \mid |x| \le M, |y| \le M\}$.

Of course, the family of sets $\{R^3_M \mid M > 0\}$ can be replaced by any family $\{K_M \mid M > 0\}$ of compact sets whose characteristic functions tend to 1 almost everywhere in R^3 when $M \to \infty$. Corollary 9.6 implies a similar representation for the adjoint operators $\Psi^*_\pm$, Ψ^*_k. It is formulated as

Corollary 9.7. The operators $\Psi^*_\pm$, Ψ^*_k have the representations

$$(9.27) \qquad (\Psi^*_\pm g_\pm)(x,y) = \mathcal{H}\text{-}\lim_{M\to\infty} \int_{\Omega^M} \psi_\pm(x,y,p,\lambda)\, g_\pm(p,\lambda)dpd\lambda,$$

$$(9.28) \qquad (\Psi^*_0 g_0)(x,y) = \mathcal{H}\text{-}\lim_{M\to\infty} \int_{\Omega^M_0} \psi_0(x,y,p,\lambda)\, g_0(p,\lambda)dpd\lambda,$$

$$(9.29) \qquad (\Psi^*_k g_k)(x,y) = \mathcal{H}\text{-}\lim_{M\to\infty} \int_{\Omega^M_k} \psi_k(x,y,p)\, g_k(p)dp,$$

for $1 \le k < N_0$ where Ω^M and Ω^M_k $(0 \le k < N_0)$ are families of compact subsets of Ω and Ω_k whose characteristic functions tend to 1 almost everywhere in Ω and Ω_k, respectively, when $M \to \infty$.

By combining Theorem 9.1 and Corollaries 9.2, 9.6 and 9.7 the following explicit formulation of the normal mode expansion is obtained.

Theorem 9.8. Every $f \in \mathcal{H}$ has a representation

$$(9.30) \qquad f(x,y) = f_+(x,y) + f_-(x,y) + \sum_{k=0}^{N_0-1} f_k(x,y),$$

convergent in $\mathcal{H}$, where $f_\pm \in \mathcal{H}_\pm$ and f_k, $0 \le k < N_0$, are given by

$$(9.31) \qquad f_\pm(x,y) = \mathcal{H}\text{-}\lim_{M\to\infty} \int_{\Omega^M} \psi_\pm(x,y,p,\lambda)\, \tilde{f}_\pm(p,\lambda)dpd\lambda,$$

$$(9.32) \qquad f_0(x,y) = \mathcal{H}\text{-}\lim_{M\to\infty} \int_{\Omega^M_0} \psi_0(x,y,p,\lambda)\, \tilde{f}_0(p,\lambda)dpd\lambda,$$

$$(9.33) \qquad f_k(x,y) = \mathcal{H}\text{-}\lim_{M\to\infty} \int_{\Omega^M_k} \psi_k(x,y,p)\, \tilde{f}_k(p)dp,$$

for $1 \le k < N_0$ and $\tilde{f}_\pm$, $\tilde{f}_k$ are defined by (8.14)-(8.16). Conversely, if $\tilde{f} = (\tilde{f}_+,\tilde{f}_-,\tilde{f}_0,\tilde{f}_1,\cdots)$ is any vector in $\tilde{\mathcal{H}}$ then (9.30)-(9.33) define a vector $f \in \mathcal{H}$ such that $\tilde{f}_\pm$, $\tilde{f}_k$ are related to f by (8.14)-(8.16).

Theorem 9.9. The unitary operator Ψ defines a spectral representation for A in $\tilde{\mathcal{H}}$ in the sense that for all $f \in D(A)$ one has

$$(9.34) \qquad (\Psi_\pm A\, f)(p,\lambda) = \lambda(\Psi_\pm f)(p,\lambda) = \lambda\, \tilde{f}_\pm(p,\lambda),$$

$$(9.35)\qquad (\Psi_0 A\, f)(p,\lambda) = \lambda(\Psi_0 f)(p,\lambda) = \lambda\, \tilde{f}_0(p,\lambda),$$

$$(9.36)\qquad (\Psi_k A\, f)(p) = \lambda_k(|p|)(\Psi_k f)(p) = \lambda_k(|p|)\, \tilde{f}_k(p),$$

for $1 \le k < N_0$.

Corollary 9.10. The complete family of orthogonal projections $\{P_+,P_-,P_0,P_1,\cdots\}$ reduces A; i.e.,

$$(9.37)\qquad P_\pm A \subset A\, P_\pm,\ P_k A \subset A\, P_k$$

for $0 \le k < N_0$ and if

$$(9.38)\qquad \begin{aligned} A_\pm &= A\, P_\pm = P_\pm\, A\, P_\pm \\ A_k &= A\, P_k = P_k\, A\, P_k,\ 0 \le k < N_0, \end{aligned}$$

denote the parts of A in $\mathcal{H}_\pm$ and $\mathcal{H}_k$ then

$$(9.39)\qquad A = A_+ + A_- + \sum_{k=0}^{N_0-1} A_k.$$

The Family $\{\phi_+(x,y,p,q) \mid (p,q) \in R^3 - N\}$. It will now be shown how the normal mode expansion of Theorem 9.8 can be reformulated in terms of the family $\{\phi_+,\psi_1,\psi_2,\cdots\}$. To begin, equation (1.44) for the normalizing factor $c(p,q)$ will be verified.

First, recall that the normalizing factors $a_\pm(\mu,\lambda)$, $a_0(\mu,\lambda)$ for $\psi_\pm$, ψ_0 are related to the factors $c_\pm(\mu,\lambda)$, $c_0(\mu,\lambda)$ in their asymptotic forms, equations (1.18)-(1.20), by

$$(9.40)\qquad c_\pm = T_\pm^{-1}\, a_\pm,\ c_0 = T_0^{-1}\, a_0;$$

see (5.8), (5.14), (5.21). Combining (9.40) with equations (5.10), (5.16), (5.23) for $T_\pm$, T_0 and equations (6.10), (6.14), (6.19), (6.21) defining $a_\pm$, a_0 gives

$$(9.41)\qquad c_\pm(\mu,\lambda) = \left(\frac{\rho(\pm\infty)}{4\pi\, q_\pm(\mu,\lambda)}\right)^{1/2},\ c_0(\mu,\lambda) = \left(\frac{p(\infty)}{4\pi\, q_+(\mu,\lambda)}\right)^{1/2},$$

provided that the phase factors $e^{i\theta_\pm}$, $e^{i\theta_0}$ are defined by

$$(9.42)\qquad \begin{aligned} e^{i\theta_\pm(\mu,\lambda)} &= \frac{i[\phi_3\phi_2]}{|[\phi_3\phi_2]|}, \ \lambda \in \Lambda(\mu), \\ e^{i\theta_0(\mu,\lambda)} &= \frac{i[\phi_3\phi_2]}{|[\phi_3\phi_2]|}, \ \lambda \in \Lambda_0(\mu). \end{aligned}$$

On combining (9.41) and the definition of $\phi_+(x,y,p,q)$, equations (1.38) and (1.39), and the asymptotic forms (1.18)-(1.20) for $\psi_\pm$, ψ_0 one obtains the asymptotic forms (1.42), (1.43) for ϕ_+ with the normalizing factor $c(p,q)$ defined by (1.44).

To derive the normal mode expansion for the family $\{\phi_+,\psi_1,\psi_2,\cdots\}$ from that for $\{\psi_+,\psi_-,\psi_0,\psi_1,\psi_2,\cdots\}$ it is clearly sufficient to restrict attention to the subspace

$$(9.43)\qquad \mathcal{H}_f = \mathcal{H}_+ + \mathcal{H}_- + \mathcal{H}_0$$

and the corresponding orthogonal projection

$$(9.44)\qquad P_f = P_+ + P_- + P_0 .$$

Thus if $h \in \mathcal{H}$ and $h_f = P_f h = h_+ + h_- + h_0$ then Theorem 9.8 implies that

$$(9.45)\qquad \begin{aligned} h_f(x,y) = &\int_\Omega \psi_+(x,y,p,\lambda)\ \tilde{h}_+(p,\lambda)dpd\lambda + \int_{\Omega_0} \psi_0(x,y,p,\lambda)\ \tilde{h}_0(p,\lambda)dpd\lambda \\ &+ \int_\Omega \psi_-(x,y,p,\lambda)\ \tilde{h}_-(p,\lambda)dpd\lambda \end{aligned}$$

where the integrals converge in $\mathcal{H}$. Changing the variables of integration in the three integrals by means of the mappings X_+, X_0 and X_-, respectively, and using the definition (1.38), (1.39) of ϕ_+ gives

$$(9.46)\qquad \begin{aligned} h_f(x,y) &= \int_{C_+} \phi_+(x,y,p,q)\ \tilde{h}_+(p,\lambda)\ c(\infty)(2q)^{1/2}\ dpdq \\ &+ \int_{C_0} \phi_+(x,y,p,q)\ \tilde{h}_0(p,\lambda)\ c(\infty)(2q)^{1/2}\ dpdq \\ &+ \int_{C_-} \phi_+(x,y,p,q)\ \tilde{h}_-(p,\lambda)\ c(-\infty)(2|q|)^{1/2}\ dpdq \\ &= \int_{R^3} \phi_+(x,y,p,q)\ \hat{h}_+(p,q)dpdq \end{aligned}$$

where $\lambda = \lambda(p,q)$ is defined by (1.41) and

$$\hat{h}_+(p,q) = \begin{cases} (2q)^{1/2}\, c(\infty)\, \tilde{h}_+(p,\lambda(p,q)), & (p,q) \in C_+, \\ (2q)^{1/2}\, c(\infty)\, \tilde{h}_0(p,\lambda(p,q)), & (p,q) \in C_0, \\ (2|q|)^{1/2}\, c(-\infty)\, \tilde{h}_-(p,\lambda(p,q)), & (p,q) \in C_-, \\ 0, & (p,q) \in N. \end{cases} \tag{9.47}$$

It is easy to verify by considering the three cones C_+, C_0 and C_- separately that

$$\hat{h}_+(p,q) = \int_{R^3} \overline{\phi_+(x,y,p,q)}\, h(x,y)\, c^{-2}(y)\, \rho^{-1}(y)\, dx dy \tag{9.48}$$

where the integral converges in $L_2(R^3)$. Moreover, it can be verified by direct calculation, using the Parseval relation of Theorem 8.3, that

$$\|h_f\|_{\mathcal{H}} = \|\hat{h}_+\|_{L_2(R^3)} \tag{9.49}$$

for all $h \in \mathcal{H}$. These considerations suggest

Theorem 9.11. For all $h \in \mathcal{H}$ the limit

$$\hat{h}_+(p,q) = L_2(R^3)\text{-}\lim_{M\to\infty} \int_{R^3_M} \overline{\phi_+(x,y,p,q)}\, h(x,y) c^{-2}(y) \rho^{-1}(y)\, dx dy \tag{9.50}$$

exists. Moreover, the mapping $\Phi_+ : \mathcal{H} \to L_2(R^3)$ defined by $\Phi_+ h = \hat{h}_+$ is a partial isometry such that

$$\Phi_+ \Phi_+^* = 1 \text{ and } \Phi_+^* \Phi_+ = P_f, \tag{9.51}$$

and the adjoint mapping $h_f = \Phi_+^* \hat{h}_+$ is given by

$$h_f(x,y) = \mathcal{H}\text{-}\lim_{M\to\infty} \int_{(R^3-N)_M} \phi_+(x,y,p,q)\, \hat{h}_+(p,q)\, dp dq. \tag{9.52}$$

Finally, Φ_+ is a spectral mapping for A in the sense that for all $h \in D(A)$ one has

$$(\Phi_+ A\, h)(p,q) = \lambda(p,q)\, \Phi_+ h(p,q) \tag{9.53}$$

where $\lambda(p,q)$ is defined by (1.41).

Note that Theorem 9.11 is simply a reformulation of Theorem 9.8 and not a new theorem.

The Family $\{\phi_-(x,y,p,q) \mid (p,q) \in R^3 - N\}$. The analogue of Theorem 9.11 for the family $\{\phi_-\}$ will be formulated as

Corollary 9.12. For all $h \in \mathcal{H}$ the limit

$$(9.54) \qquad \hat{h}_-(p,q) = L_2(R^3)\text{-}\lim_{M\to\infty} \int_{R_M^3} \overline{\phi_-(x,y,p,q)}\, h(x,y) c^{-2}(y) \rho^{-1}(y)\, dx dy$$

exists and the mapping $\Phi_- : \mathcal{H} \to L_2(R^3)$ defined by $\Phi_- h = \hat{h}_-$ is a partial isometry such that

$$(9.55) \qquad \Phi_- \Phi_-^* = 1 \text{ and } \Phi_-^* \Phi_- = P_f.$$

Moreover,

$$(9.56) \qquad h_f(x,y) = \mathcal{H}\text{-}\lim_{M\to\infty} \int_{R_M^3} \phi_-(x,y,p,q)\, \hat{h}_-(p,q)\, dp dq.$$

Finally, for all $h \in D(A)$, one has

$$(9.57) \qquad (\Phi_- A h)(p,q) = \lambda(p,q)\, \Phi_- h(p,q).$$

These results are direct corollaries of Theorem 9.11. This follows from the observations that $f(p,q) \to f(-p,q)$ defines a unitary transformation in $L_2(R^3)$ while $f \to \overline{f}$ defines a unitary transformation in both $\mathcal{H}$ and $L_2(R^3)$.

This completes the formulation of the results of §9. The proofs will now be presented.

The proofs of Theorem 9.1 and Corollaries 9.2 and 9.3 will be based on a lemma concerning bounded linear operators from a Hilbert space into a direct sum space. To formulate it let $\mathcal{H}$ and $\mathcal{H}_k$ $(k \in N)$ denote Hilbert spaces, where N is a finite or denumerable set, and define

$$(9.58) \qquad \tilde{\mathcal{H}} = \sum_{k\in N} \mathcal{H}_k.$$

Elements of $\tilde{\mathcal{H}}$ will be written $g = \{g_k\}$ where $g_k \in \mathcal{H}_k$ and

$$(9.59) \qquad \|g\|^2 = \sum_{k\in N} \|g_k\|_k^2 < \infty$$

if $\|\cdot\|_k$ is the norm in $\mathcal{H}_k$. If

$$(9.60) \qquad B : \mathcal{H} \to \tilde{\mathcal{H}}$$

is a bounded linear operator then

$$Bf = \{(Bf)_k\} = \{B_k f\} \tag{9.61}$$

where

$$B_k : \mathcal{H} \to \mathcal{H}_k \tag{9.62}$$

is a bounded linear operator. With this notation one has

<u>Lemma 9.13</u>. The adjoint operator $B^* : \tilde{\mathcal{H}} \to \mathcal{H}$ is given by

$$B^* g = \sum_{k \in N} B_k^* g_k, \quad g = \{g_k\} \in \tilde{\mathcal{H}}, \tag{9.63}$$

where the series converges in $\mathcal{H}$. Hence

$$B^* B = \sum_{k \in N} B_k^* B_k, \tag{9.64}$$

where the series converges strongly, and

$$B\, B^* g = \left\{ \sum_{j \in N} B_k B_j^* g_j \right\}, \quad g = \{g_k\} \in \tilde{\mathcal{H}}, \tag{9.65}$$

where the series converge in $\mathcal{H}_k$.

<u>Proof</u>. If N is finite then (9.63) follows directly from the definition of B^*. If N is denumerably infinite then (9.63) holds for all $g \in \tilde{\mathcal{H}}$ with finitely many non-zero components g_k. But any $g \in \tilde{\mathcal{H}}$ can be approximated in $\tilde{\mathcal{H}}$ by such vectors and B^* is bounded. The convergence in $\mathcal{H}$ of the series in (9.63) follows. Equation (9.64) follows on applying (9.63) to the vector (9.61). (9.65) follows on applying B to the vector (9.63).

<u>Corollary 9.14</u>. $B^* B = 1$ in $\mathcal{H}$ if and only if

$$\sum_{k \in N} B_k^* B_k = 1 \text{ in } \mathcal{H}, \tag{9.66}$$

the series converging strongly.

<u>Corollary 9.15</u>. $B\, B^* = 1$ in $\tilde{\mathcal{H}}$ if and only if

$$B_k B_j^* = \delta_{jk} \text{ for all } k, j \in N \tag{9.67}$$

where $\delta_{jk} = 1$ if $j = k$ and $\delta_{jk} = 0$ if $j \neq k$.

Proof of Corollaries 9.14 and 9.15. (9.66) is an immediate consequence of (9.64) and $B^*B = 1$. (9.67) follows from $B\,B^* = 1$ and (9.65) on taking $g_{(j)} = \{\delta_{jk}\, g_k\}$, j fixed.

On applying Corollary 9.14 to the direct sum space (9.1) and the operator Ψ it is seen that (9.13) of Corollary 9.2 is equivalent to (9.5) of Theorem 9.1. Similarly, Corollary 9.15 implies that the relations of Corollary 9.3 are equivalent to (9.6) of Theorem 9.1.

Proof of Corollary 9.2. It was shown that (9.5) follows from Theorem 8.3. Thus (9.13) is valid by Corollary 9.14.

Proof of Corollary 9.3. The spectral mapping $\Psi_\mu = (\Psi_{\mu+}, \Psi_{\mu-}, \Psi_{\mu 0}, \Psi_{\mu 1}, \cdots)$ for A_μ is unitary for all $\mu > 0$ by Theorem 6.6. It follows from Corollary 9.15 that the analogue of Corollary 9.3 holds for Ψ_μ; see (6.38). It will be shown that Corollary 9.3 follows from these relations. For brevity only the relation $\Psi_0\, \Psi_0^* = 1$ will be proved. The remaining relations can be proved by the same method.

For the proof of $\Psi_0\, \Psi_0^* = 1$ define

$$\hat{\Psi}_0 = \Psi_0\, F^{-1} : \mathcal{H} \to L_2(\Omega_0). \tag{9.68}$$

Then

$$\Psi_0\, \Psi_0^* = \hat{\Psi}_0\, \hat{\Psi}_0^* \tag{9.69}$$

and it will suffice to prove that $\hat{\Psi}_0\, \hat{\Psi}_0^* = 1$. The following lemma will be used.

Lemma 9.16. For all $g \in \mathcal{D}(\Omega_0)$ one has

$$(\hat{\Psi}_0^*\, g)(p,y) = \int_{\Lambda_0(|p|)} \psi_0(y,|p|,\lambda)\; g(p,\lambda)\,d\lambda = [\Psi_{|p|0}^*\; g(p,\cdot)](y). \tag{9.70}$$

Proof of Lemma 9.16. Let $f \in \mathcal{H}' = F^{-1}\, \mathcal{D}(R^3)$. Then $\hat{f} \in \mathcal{D}(R^3)$ and by Lemma 8.10 one has

$$\begin{aligned}(\hat{\Psi}_0^* g, \hat{f}) &= (g, \hat{\Psi}_0 \hat{f}) = (g, \Psi_0 f) = (g, \tilde{f}_0) \\ &= \int_{\Omega_0} \overline{g(p,\lambda)} \int_R \overline{\psi_0(y,|p|,\lambda)}\; \hat{f}(p,y) c^{-2}(y) \rho^{-1}(y)\, dp\, d\lambda.\end{aligned} \tag{9.71}$$

Now $\psi_0(y,|p|,\lambda)$ is defined for $(y,p,\lambda) \in R \times \Omega_0$. Extend the definition to all $(y,p,\lambda) \in R \times R^3$ by

(9.72) $$\psi_0(y,|p|,\lambda) = 0 \text{ for } y \in R,\ (p,\lambda) \in R^3 - \Omega_0,$$

and apply Fubini's theorem to the integral in (9.71). This gives

(9.73) $$(\hat{\Psi}_0^* g, \hat{f}) = \int_{R^3} \left[\int_R \overline{\psi_0(y,|p|,\lambda) g(p,\lambda)}\, d\lambda \right] \hat{f}(p,y) c^{-2}(y) \rho^{-1}(y)\, dp dy,$$

which implies (9.70) because $\mathcal{D}(R^3)$ is dense in $\mathcal{H}$ and supp $\psi_0(y,|p|,\cdot) \subset \Lambda_0(|p|)$, by (9.72).

Proof of Corollary 9.3 (completed). The relation $\hat{\Psi}_0\ \hat{\Psi}_0^* = 1$ is equivalent to

(9.74) $$(\hat{\Psi}_0\ \hat{\Psi}_0^* f, g) = (\hat{\Psi}_0^* f, \hat{\Psi}_0^* g) = (f,g)$$

for all $f,g \in L_2(\Omega_0)$. Moreover, since $\hat{\Psi}_0^*$ is bounded it will suffice to verify (9.74) for f and g in the dense set $\mathcal{D}(\Omega_0)$. Now Lemma 9.16 implies that for all $f,g \in \mathcal{D}(\Omega_0)$ one has

(9.75) $$\begin{aligned} (\hat{\Psi}_0^* f, \hat{\Psi}_0^* g) &= \int_{R^3} \overline{[\Psi_{|p|0}^* f(p,\cdot)]\ (y)}\ [\Psi_{|p|0}^* g(p,\cdot)](y)\, dp dy \\ &= \int_{R^2} (\Psi_{|p|0}^* f(p,\cdot), \Psi_{|p|0}^* g(p,\cdot))\, dp. \end{aligned}$$

But by (6.38), $\Psi_{|p|0}\ \Psi_{|p|0}^* = 1$ for all $|p| > 0$ and hence the last integral equals

(9.76) $$\int_{R^2} (f(p,\cdot), g(p,\cdot))_{\Lambda_0(|p|)}\, dp = (f,g)_{L_2(\Omega_0)}$$

which proves (9.74). This completes the proof.

Proof of Theorem 9.1. It was shown that (9.5) follows from Theorem 8.3. Relation (9.6) follows from Corollary 9.3, by Corollary 9.15.

Proof of Corollaries 9.4 and 9.5. Corollary 9.4 follows immediately from the definitions (9.16) and Corollaries 9.2 and 9.3. Corollary 9.5 follows from Corollary 9.4.

Proof of Corollary 9.6. This is just a restatement of Theorem 8.3, equation (8.12).

Proof of Corollary 9.7. Equation (8.27) for Ψ_+^* will be verified. The proofs of the remaining equations are similar. It is clearly sufficient to verify (9.27) for functions $g_+ \in L_2(\Omega)$ with compact supp $g_+ \subset \Omega^M$. If g_+ is such a function and $f \in \mathcal{H}^{com}$ then

$$
\begin{aligned}
(f,\Psi_+^* g) &= (\Psi_+ f, g_+) = (\tilde{f}_+, g_+) \\
&= \int_{\Omega^M} \left[\int_{R^3} \psi_+(x,y,p,\lambda)\, \overline{f(x,y)}\, c^{-2}(y)\rho^{-1}(y)\,dxdy \right] g_+(p,\lambda)\,dpd\lambda \\
&= \int_{R^3} \overline{f(x,y)} \left[\int_{\Omega^M} \psi_+(x,y,p,\lambda) g_+(p,\lambda)\,dpd\lambda \right] c^{-2}(y)\rho^{-1}(y)\,dxdy.
\end{aligned} \tag{9.77}
$$

This relation implies (9.27) for g_+ because $\mathcal{H}^{com}$ is dense in $\mathcal{H}$.

Proof of Theorem 9.8. The representation (9.30), (9.31) follows from Corollaries 9.2 and 9.7. The converse follows from the unitarity of Ψ; Theorem 9.1.

Proof of Theorem 9.9. Only equation (9.35) will be proved. The other equations can be proved by the same method. To prove (9.35) let $f \in D(A)$, $g \in \mathcal{D}(\Omega_0)$ and note that

$$
(\Psi_0 A\, f, g) = (A\, f, \Psi_0^* g) = (\hat{A}\, \hat{f}, \hat{\Psi}_0^* g). \tag{9.78}
$$

Now if $h = \hat{\Psi}_0^* g$ then $h \in \mathcal{H}$, by Corollary 9.3 and, by Lemma 9.16,

$$
h(p,y) = \int_{\Lambda_0(|p|)} \psi_0(y,|p|,\lambda)\, g(p,\lambda)\,d\lambda, \quad (p,y) \in R^3. \tag{9.79}
$$

A distribution-theoretic calculation of $\hat{A}\, h$ gives

$$
\hat{A}\, h(p,y) = \int_{\Lambda_0(|p|)} \psi_0(y,|p|,\lambda)\, \lambda\, g(p,\lambda)\,d\lambda. \tag{9.80}
$$

In particular, since $\lambda\, g(p,\lambda) \in L_2(\Omega_0)$, one has $\hat{A}\, h \in \mathcal{H}$ by Corollary 9.3 and hence $h \in D(\hat{A})$. Combining this with (9.78) gives

$$
\begin{aligned}
(\Psi_0 A\, f, g) &= (\hat{A}\, \hat{f}, h) = (\hat{f}, \hat{A}\, h) \\
&= \int_{R^3} \overline{\hat{f}(p,y)} \left[\int_{\Lambda_0(|p|)} \psi_0(y,|p|,\lambda)\lambda g(p,\lambda)\,d\lambda \right] c^{-2}(y)\rho^{-1}(y)\,dpdy.
\end{aligned} \tag{9.81}
$$

Writing $k(p,\lambda) = \lambda\, g(p,\lambda) \in \mathcal{D}(\Omega_0)$ and applying Lemma 9.16 again gives

$$
\begin{aligned}
(\Psi_0 A\, f, g) &= (\hat{f}, \hat{\Psi}_0^* k) = (f, \Psi_0^* k) = (\Psi_0 f, k) \\
&= \int_{\Omega} \overline{\tilde{f}_0(p,\lambda)}\, k(p,\lambda)\,dpd\lambda = \int_{\Omega} \overline{\lambda\, \tilde{f}_0(p,\lambda)}\, g(p,\lambda)\,dpd\lambda.
\end{aligned} \tag{9.82}
$$

This implies (9.35) because the functions $g \in \mathcal{D}(\Omega_0)$ are dense in $L_2(\Omega_0)$.

Proof of Corollary 9.10. It is only necessary to verify relations (9.37). By [11, p. 530] these relations are equivalent to

$$(9.83) \qquad P_{\pm}\Pi(\mu) = \Pi(\mu)P_{\pm},\ P_k\Pi(\mu) = \Pi(\mu)P_k$$

for $0 \le k < N_0$ and all $\mu \in R$. These equations are immediate consequences of Theorem 8.4. For example (8.17) and the definition of P_+ implies that

$$(9.84) \qquad (f,P_+\Pi(\mu)g) = (f,\Pi(\mu)P_+g) = \int_\Omega H(\mu-\lambda)\ \overline{\tilde{f}_+(p,\lambda)}\ \tilde{g}_+(p,\lambda)dpd\lambda$$

for all $f,g \in \mathcal{H}$, which implies $P_+\Pi(\mu) = \Pi(\mu)P_+$. The other relations are proved similarly.

Proof of Theorem 9.11. Equation (9.50) can be verified by applying the definition (9.47) to h_M and using the convergence statement of Theorem 8.3. Relations (9.51) follow from (9.47) and Corollary 9.3 by direct calculation. Equation (9.52) can be verified by reversing the steps in the calculation (9.45), (9.46). Relation (9.53) follows from Theorem 9.9 and equation (9.47).

Proof of Corollary 9.12. This was indicated immediately after the statement of the Corollary.

§10. SEMI-INFINITE AND FINITE LAYERS

The purpose of this section is to present extensions of the preceding analysis to the cases of semi-infinite and finite layers of stratified fluid. The methods and results are entirely analogous to those developed above. Therefore the presentation emphasizes the modifications required in these cases. Proofs are indicated briefly or omitted.

Semi-Infinite Layers. With a suitable choice of coordinates the region occupied by the fluid is described by the domain

$$(10.1) \qquad R^3_+ = \{(x,y) \mid y > 0\}.$$

The acoustic field is assumed to satisfy either the Dirichlet or the Neumann boundary condition on the boundary of R^3_+. Physically, these conditions correspond to the cases where the boundary plane is free and rigid, respectively. The functions $\rho(y)$ and $c(y)$ are assumed to be Lebesgue measurable and satisfy

$$(10.2) \qquad 0 < \rho_m \le \rho(y) \le \rho_M < \infty,\ 0 < c_m \le c(y) \le c_M < \infty$$

for all $y > 0$ and

$$(10.3) \qquad \int_0^\infty |\rho(y) - \rho(\infty)|dy < \infty, \quad \int_0^\infty |c(y) - c(\infty)|dy < \infty.$$

The acoustic propagator A defined by (2.1.12) determines selfadjoint operators A^0 and A^1 in

$$(10.4) \qquad \mathcal{H}_+ = L_2(R_+^3, c^{-2}(y)\rho^{-1}(y)dxdy)$$

corresponding to the two boundary conditions. The domains of A^0 and A^1 are subsets of

$$(10.5) \qquad L_2^1(A, R_+^3) = L_2^1(R_+^3) \cap \{u \mid \nabla \cdot (\rho^{-1}\nabla u) \in L_2(R_+^3)\}.$$

The operator A^0 corresponding to the Dirichlet condition is defined by

$$(10.6) \qquad D(A^0) = L_2^1(A, R_+^3) \cap \{u \mid u(x,0+) = 0 \text{ in } L_2(R^2)\}.$$

Sobolev's embedding theorem for $L_2^1(R_+^3)$ [1] implies that the boundary values $u(x,0+)$ are defined in $L_2(R^2)$.

The Neumann condition will be interpreted in the generalized sense employed in [22]; i.e.,

$$(10.7) \qquad \int_{R_+^3} \{\nabla \cdot (\rho^{-1}\nabla u)v + \rho^{-1}\nabla u \cdot \nabla v\}dxdy = 0$$

for all $v \in L_2^1(R_+^3)$. Thus the operator A^1 corresponding to the Neumann condition is defined by

$$(10.8) \qquad D(A^1) = L_2^1(A, R_+^3) \cap \{u \mid (10.7) \text{ holds for all } v \in L_2^1(R_+^3)\}.$$

The operators are defined by $A^j u = Au$ for all $u \in D(A^j)$ and one has

$$(10.9) \qquad A^j = A^{j*} \geq 0, \; j = 0, 1.$$

This is most easily proved by introducing the corresponding sesquilinear forms, as in Chapter 2, and using Kato's representation theorem [11, p. 322].

The spectral analysis of A^0 and A^1 may be based on the corresponding reduced propagators A_μ^0 and A_μ^1 in

$$(10.10) \qquad \mathcal{H}(R_+) = L_2(R_+, c^{-2}(y)\rho^{-1}(y)dy), \; R_+ = \{y \mid y > 0\}.$$

They are defined by

(10.11) $$D(A_\mu^0) = L_2^1(R_+) \cap \{\psi \mid (\rho^{-1}\psi')' \in \mathcal{H}(R_+) \text{ and } \psi(0+) = 0\},$$

(10.12) $$D(A_\mu^1) = L_2^1(R_+) \cap \{\psi \mid (\rho^{-1}\psi')' \in \mathcal{H}(R_+) \text{ and } (\rho\psi')(0+) = 0\},$$

(10.13) $$A_\mu^j\psi = A_\mu\psi \text{ for } \psi \in D(A_\mu^j),\ j = 0,1,$$

and one has

(10.14) $$A_\mu^j = A_\mu^{j*} \geq c_m^2\,\mu^2,\ j = 0,1.$$

The results of §4 can be extended to these operators. Thus

(10.15) $$\sigma_c(A_\mu^j) = \sigma_e(A_\mu^j) = [c^2(\infty)\mu^2,\infty),$$

(10.16) $$\sigma(A_\mu^j) \cap (-\infty, c^2(\infty)\mu^2) \subset \sigma_0(A_\mu^j),$$

and the eigenvalues in this interval are all simple. They will be denoted by $\lambda_k^j(\mu)$, $1 \leq k < N^j(\mu) \leq +\infty$.

Eigenfunctions of A_μ^j. These functions will be denoted by $\psi_k^j(y,\mu)$, $1 \leq k < N^j(\mu)$. They are uniquely defined by the conditions $\psi_k^j(\cdot,\mu) \in D(A_\mu^j)$, $\|\psi_k^j(\cdot,\mu)\|_{\mathcal{H}(R_+)} = 1$,

(10.17) $$(A_\mu - \lambda_k^j(\mu))\,\psi_k^j(y,\mu) = 0 \text{ for } y \in R_+, \text{ and}$$

(10.18) $$\psi_k^0(0+,\mu) = 0,\ (\rho^{-1}\psi_k^{1\prime})(0+,y) = 0.$$

The asymptotic behavior of $\psi_k^j(y,\mu)$ for $y \to +\infty$ is given by

(10.19) $$\psi_k^j(y,\mu) \sim c_k^j(\mu) \exp\{-y\, q'(\mu,\lambda_k^j(\mu))\},\ y \to \infty,$$

where

(10.20) $$q'(\mu,\lambda) = (\mu^2 - \lambda\, c^{-2}(\infty))^{1/2} > 0.$$

The function $\psi_k^j(y,\mu)$ has precisely $k - 1$ zeros.

Generalized Eigenfunctions of A_μ^j. For $\lambda > c^2(\infty)\mu^2$, A_μ^j has a single family of generalized eigenfunctions $\{\psi^j(\cdot,\mu,\lambda) \mid \lambda > c^2(\infty)\mu^2\}$. They are determined up to normalization by the conditions

$$(A_\mu - \lambda)\psi^j(y,\mu,\lambda) = 0 \text{ for } y \in R_+, \text{ and} \tag{10.21}$$

$$\psi^0(0+,\mu,\lambda) = 0, \quad (\rho^{-1}\psi^{1\prime})(0+,\mu,\lambda) = 0. \tag{10.22}$$

Their asymptotic behavior for $y \to \infty$ is given by

$$\psi^j(y,\mu,\lambda) \sim c^j(\mu,\lambda)[e^{-iyq(\mu,\lambda)} + R^j(\mu,\lambda)\, e^{iyq(\mu,\lambda)}], \quad y \to \infty, \tag{10.23}$$

where

$$q(\mu,\lambda) = (\lambda\, c^{-2}(\infty) - \mu^2)^{1/2} > 0. \tag{10.24}$$

The normalizing factors $c^j(\mu,\lambda)$ are calculated below.

Generalized Eigenfunctions of A^j. These are defined by

$$\psi^j(x,y,p,\lambda) = (2\pi)^{-1}\, e^{ip\cdot x}\, \psi^j(y,|p|,\lambda), \quad (p,\lambda) \in \Omega, \tag{10.25}$$

$$\psi_k^j(x,y,p) = (2\pi)^{-1}\, e^{ip\cdot x}\, \psi_k^j(y,|p|), \quad p \in \Omega_k, \quad k \geq 1 \tag{10.26}$$

where

$$\Omega = \{(p,\lambda) \mid \lambda > c^2(\infty)|p|^2\} \tag{10.27}$$

$$\Omega_k = \{p \mid |p| \in \mathcal{O}_k\}, \quad k \geq 1 \tag{10.28}$$

and $\mathcal{O}_k = \{\mu \mid N^j(\mu) \geq k + 1\}$, as before. The wave-theoretic interpretations of the eigenfunctions (10.25), (10.26) may be derived from the asymptotic forms (10.19), (10.23) as in §1. With these definitions the following analogue of Theorem 8.4 holds.

Theorem 10.1. The spectral families $\{\Pi^j(\mu)\}$ of A^j satisfy

$$\begin{aligned}(f,\Pi^j(\mu)f) &= \int_\Omega H(\mu-\lambda)\, |\tilde{f}^j(p,\lambda)|^2\, dp d\lambda \\ &+ \sum_{k=1}^{N_0^j-1} \int_{\Omega_k} H(\mu - \lambda_k^j(|p|))\, |\tilde{f}_k^j(p)|^2\, dp\end{aligned} \tag{10.29}$$

for all $f \in \mathcal{H}_+$ where $N_0^j = \sup_{\mu>0} N^j(\mu)$ and

$$(10.30) \qquad \tilde{f}^j(p,\lambda) = L_2(\Omega)\text{-}\lim_{M\to\infty} \int_0^M \int_{|x|\le M} \overline{\psi^j(x,y,p,\lambda)}\, f(x,y)c^{-2}(y)\rho^{-1}(y)\,dxdy,$$

$$(10.31) \qquad \tilde{f}_k^j(p) = L_2(\Omega_k)\text{-}\lim_{M\to\infty} \int_0^M \int_{|x|\le M} \overline{\psi_k^j(x,y,p)}\, f(x,y)c^{-2}(y)\rho^{-1}(y)\,dxdy.$$

Relation to the Infinite Layer Problem. Theorem 10.1 can be derived by the method employed for the infinite layer problem in the preceding sections. However, it can also be deduced directly from Theorem 8.4. To this end one extends $\rho(y)$, $c(y)$ to all $y \in R$ as even functions:

$$(10.32) \qquad \rho(-y) = \rho(y),\ c(-y) = c(y),\ y > 0.$$

Then it follows from (10.2), (10.3) that the extended functions satisfy (2.1.4), (1.1) with

$$(10.33) \qquad \rho(-\infty) = \rho(\infty),\ c(-\infty) = c(\infty).$$

The corresponding operator in $\mathcal{H}$ will be denoted by A, as before. Property (10.33) implies that

$$(10.34) \qquad q_\pm(\mu,\lambda) = q(\mu,\lambda),\ q'_\pm(\mu,\lambda) = q'(\mu,\lambda)$$

where the latter are defined by (10.20), (10.24). Moreover, the special solutions $\phi_j(y,\mu,\lambda)$ of §3 satisfy

$$(10.35) \qquad \begin{cases} \phi_1(-y,\mu,\lambda) = \phi_4(y,\mu,\lambda) \\ \phi_2(-y,\mu,\lambda) = \phi_3(y,\mu,\lambda) \end{cases}$$

and it is easy to verify that the eigenfunction $\psi_k(y,\mu)$ of A_μ is even (resp., odd) when k is odd (resp., even). It follows that the eigenfunctions of A_μ^j can be calculated from those of A_μ by the rule

$$(10.36) \qquad \begin{cases} \psi_k^0(y,\mu) = \sqrt{2}\,\psi_{2k}(y,\mu), & y \ge 0,\ k = 1,2,\cdots \\ \psi_k^1(y,\mu) = \sqrt{2}\,\psi_{2k-1}(y,\mu), & y \ge 0,\ k = 1,2,\cdots \end{cases}$$

The factor $\sqrt{2}$ is to renormalize ψ_k from R to R_+.

Concerning the generalized eigenfunctions, note that there is no $\psi_0(y,\mu,\lambda)$ for A_μ because $c(-\infty) = c(\infty)$ and (10.35) implies that

$$\psi_\pm(-y,\mu,\lambda) = \psi_\mp(y,\mu,\lambda). \tag{10.37}$$

It follows that the coefficients $R_\pm$, $T_\pm$ in (1.19), (1.20) satisfy

$$R_+(\mu,\lambda) = R_-(\mu,\lambda),\ T_+(\mu,\lambda) = T_-(\mu,\lambda) \tag{10.38}$$

and the generalized eigenfunctions of A_μ^j can be calculated from those of A_μ by the rule

$$\begin{aligned} \psi^0(y,\mu,\lambda) &= \psi_+(y,\mu,\lambda) - \psi_+(-y,\mu,\lambda),\ y \geq 0, \\ \psi^1(y,\mu,\lambda) &= \psi_+(y,\mu,\lambda) + \psi_+(-y,\mu,\lambda),\ y \geq 0. \end{aligned} \tag{10.39}$$

The factors $c^j(\mu,\lambda)$ and $R^j(\mu,\lambda)$ of (10.23) are given by

$$c^0(\mu,\lambda) = c^1(\mu,\lambda) = \left(\frac{\rho(\infty)}{4\pi q(\mu,\lambda)}\right)^{1/2}, \tag{10.40}$$

$$R^0 = R_\pm - T_\pm,\ R^1 = R_\pm + T_\pm. \tag{10.41}$$

Theorem 10.1 can be deduced from Theorem 8.4 by introducing the operators

$$J_j : \mathcal{H}(R_+^3) \to \mathcal{H},\ j = 0,1, \tag{10.42}$$

defined by

$$J_0 u(x,y) = \begin{cases} u(x,y), & (x,y) \in R_+^3, \\ -u(x,-y), & (x,-y) \in R_+^3, \end{cases} \tag{10.43}$$

and

$$J_1 u(x,y) = \begin{cases} u(x,y), & (x,y) \in R_+^3, \\ u(x,-y), & (x,-y) \in R_+^3. \end{cases} \tag{10.44}$$

J_0 and J_1 are bounded linear operators and using the fact that A has coefficients satisfying (10.32) one can show that the resolvents of A^j and A are related by

$$R(A^j,\zeta) = \frac{1}{2} J_j^* R(A,\zeta) J_j, \; j = 0,1. \tag{10.45}$$

From this and Stone's theorem (8.64) it follows that

$$\Pi^j(\mu) = \frac{1}{2} J_j^* \Pi(\mu) J_j, \; j = 0,1. \tag{10.46}$$

Theorem 10.1 follows directly from these relations and Theorem 8.4.

Finite Layers. In this case, with a suitable choice of coordinates the region occupied by the fluid is described by the domain

$$R_h^3 = R^3 \cap \{(x,y) \mid 0 < y < h\} \tag{10.47}$$

where $h > 0$. The case of a fluid layer with a free surface at $y = 0$ and a rigid bottom at $y = h$ will be discussed.

The acoustic propagator A and boundary conditions determine a self-adjoint operator A^h in

$$\mathcal{H}_h = L_2(R_h^3, c^{-2}(y)\rho^{-1}(y)dxdy). \tag{10.48}$$

To define the domain of A^h let

$$L_2^{1,0}(R_h^3) = L_2^1(R_h^3) \cap \{u \mid u(x,0+) = 0 \text{ in } L_2(R_h^3)\}. \tag{10.49}$$

The Dirichlet condition at $y = 0$ will be enforced by requiring $D(A^h) \subset L_2^{1,0}(R_h^3)$. The Neumann condition at $y = h$ will be interpreted in the generalized sense that

$$\int_{R_h^3} \{\nabla \cdot (\rho^{-1}\nabla u)v + \rho^{-1}\nabla u \cdot \nabla v\}dxdy = 0 \text{ for all } v \in L_2^{1,0}(R_h^3). \tag{10.50}$$

Thus

$$D(A^h) = L_2^{1,0}(R_h^3) \cap L_2^1(A,R_h^3) \cap \{u \mid (10.50) \text{ holds}\} \tag{10.51}$$

and $A^h u = Au$ for all $u \in D(A^h)$. As before

$$A^h = A^{h*} \geq 0. \tag{10.52}$$

The corresponding reduced propagator A_μ^h in $\mathcal{H}(R_h) = L_2(R_h, c^{-2}(y)\rho^{-1}(y)dy)$ $(R_h = \{y \mid 0 < y < h\})$ is defined by

$$(10.53) \quad D(A_\mu^h) = L_2^1(R_h) \cap \{\psi \mid (\rho^{-1}\psi')' \in \mathcal{H}(R_+), \psi(0+) = 0, (\rho^{-1}\psi')(h+) = 0\},$$

and $A_\mu^h\psi = A_\mu\psi$ for all $\psi \in D(A_\mu^h)$. One has

$$(10.54) \qquad A_\mu^h = A_\mu^{h*} \geq c_m^2\mu^2$$

as before. In the present case A_μ^h is a regular Sturm-Liouville operator. Hence

$$(10.55) \qquad \sigma(A_\mu^h) = \sigma_0(A_\mu^h)$$

and if $\lambda_k^h(\mu)$, $1 \leq k < \infty$, denotes the eigenvalues then $\lambda_k^h(\mu) \to \infty$ when $k \to \infty$. Note that in this case $\mathcal{O}_k = R_+$ and $\Omega_k = R^2$. If $\psi_k^h(y,\mu)$ denotes the corresponding eigenfunction and $\psi_k^h(x,y,p) = (2\pi)^{-1}e^{ip\cdot x}\psi_k^h(y,|p|)$ then the spectral family $\{\Pi^h(\mu)\}$ for A^h satisfies

$$(10.56) \qquad (f,\Pi^h(\mu)f) = \sum_{k=1}^{\infty} \int_{R^2} H(\mu - \lambda_k^h(|p|)) \, |\tilde{f}_k^h(p)|^2 \, dp$$

where

$$(10.57) \qquad \tilde{f}_k^h(p) = L_2(R^2)\text{-}\lim_{M\to\infty} \int_0^h \int_{|x|\leq M} \overline{\psi_k^h(x,y,p)} \; f(x,y)c^{-2}(y)\rho^{-1}(y)\,dxdy.$$

These results can be proved by the method developed above.

Chapter 4
Transient Sound Field Structure in Stratified Fluids

The purpose of this chapter is to analyze the structure of arbitrary sound fields with finite energy in stratified fluids. The principal results of the analysis imply that each such field is a sum of a free component, which behaves for large times like a diverging spherical wave, and a guided component which is approximately localized in regions $|y - y_j| < h_j$ where $c(y)$ has minima and propagates outward in horizontal planes like a diverging cylindrical wave.

The methods and results developed below were initiated by the author in 1973 in the special case of the Pekeris profile [23]. In 1974 the results were extended to the symmetric Epstein profile [24]. The general Epstein profile was treated in 1979 [29] using the spectral analysis of the Epstein profile due to Guillot and Wilcox [7,8]. A preliminary version of [29] was announced in 1978 [4]. The extension of these results to a large class of stratified fluids presented below is based on the general normal mode expansions of Chapter 3.

§1. SUMMARY

Throughout this chapter it is assumed that $\rho(y)$ and $c(y)$ satisfy the boundedness conditions (2.1.4) and the four conditions

$$\left\{ \begin{aligned} |\rho(y) - \rho(\pm\infty)| &\le C(\pm y)^{-\alpha} \\ |c(y) - c(\pm\infty)| &\le C(\pm y)^{-\alpha} \end{aligned} \right\} \text{ for } \pm y > 0 \tag{1.1}$$

where $\rho(+\infty)$, $\rho(-\infty)$, $c(+\infty)$, $c(-\infty)$, C and α are constants and

$$\alpha > \frac{3}{2}. \tag{1.2}$$

Note that (1.1), (1.2) imply condition (3.1.1) of the preceding chapter. It will be seen from the analysis that (1.1), (1.2) could be replaced by other order conditions at $y = \pm\infty$. It is not known whether the results obtained below are valid for the entire class of fluids studied in Chapter 3.

To begin consider a general sound field with finite energy:

$$u(t,\cdot) = (\cos t\, A^{1/2})\, f + (A^{-1/2} \sin t\, A^{1/2})\, g, \tag{1.3}$$

where f, $A^{1/2}$ f and g are in $\mathcal{H}$. For each fixed time t, $u(t,\cdot)$ has a normal mode decomposition as described by (3.1.58)-(3.1.63). Thus

$$u(t,X) = u_f(t,X) + u_g(t,X) \tag{1.4}$$

where

$$u_g(t,X) = \sum_{k=1}^{N_0-1} u_k(t,X) \tag{1.5}$$

and $N_0 - 1$ is the number of guided normal modes. Clearly, $1 \le N_0 \le +\infty$ and $u_g = 0$ if $N_0 = 1$. u_f and u_g will be called the free and guided components of u, respectively. The motivation for this terminology is provided by their behavior for $t \to \infty$ which is described next.

It will be convenient to begin with sound fields that have the complex representation (2.1.19)-(2.1.21); that is

$$u(t,X) = \mathrm{Re}\,\{v(t,X)\}, \tag{1.6}$$

$$v(t,\cdot) = e^{-itA^{1/2}} h, \text{ h and } A^{1/2} h \text{ in } \mathcal{H}. \tag{1.7}$$

The spectral property of the normal mode representation, (3.1.57) and (3.1.63), implies that

$$\begin{aligned} u_f(t,X) &= \mathrm{Re}\,\{v_f(t,X)\}, \\ u_g(t,X) &= \mathrm{Re}\,\{v_g(t,X)\}, \end{aligned} \tag{1.8}$$

where

$$v(t,X) = v_f(t,X) + v_g(t,X) \tag{1.9}$$

is the decomposition of v. Moreover, for the same reason,

$$v_f(t,\cdot) = e^{-itA^{1/2}} h_f \tag{1.10}$$

and

$$v_g(t,X) = \sum_{k=1}^{N_0-1} v_k(t,X) \tag{1.11}$$

where

$$v_k(t,\cdot) = e^{-itA^{1/2}} h_k. \tag{1.12}$$

Finally, using (3.1.57), (3.1.60), (3.1.61) and (3.1.63) one gets

$$v_f(t,X) = \int_{R^3} \phi_\pm(X,P)\ e^{-it\omega(P)}\ \hat{h}_\pm(P)dP \tag{1.13}$$

and

$$v_k(t,X) = \int_{\Omega_k} \psi_k(X,p)\ e^{-it\omega_k(|p|)}\ \tilde{h}_k(p)dp \tag{1.14}$$

where

$$\omega(P) = \lambda^{1/2}(P) = \begin{cases} c(\infty)|P| & \text{for } P \in C_+ \cup C\ , \\ c(-\infty)|P| & \text{for } P \in C_-, \\ 0 & \text{for } P \in N, \end{cases} \tag{1.15}$$

and

$$\omega_k(|p|) = \lambda_k^{1/2}(|p|) \text{ for } 1 \le k < N_0. \tag{1.16}$$

The integrals in (1.13) and (1.14) will, in general, converge only in $\mathcal{H}$. However, for brevity, the $\mathcal{H}$-lim notation is usually omitted below.

The integral representations (1.13), (1.14) provide the starting point for calculating the behavior for $t \to \infty$ of v_f and v_k. For v_f the intuitive idea is that for large t the wave $v_f(t,X)$ will have propagated into regions where $|y|$ is large and hence $\phi_\pm(X,P)$ is near its limiting forms for $y \to \pm\infty$, given by (3.1.42), (3.1.43), (3.1.48) and (3.1.49). On replacing $\phi_\pm$ in (1.13) by one of its limiting forms one obtains normal mode representations of waves in homogeneous fluids with parameters $\rho(\infty)$, $c(\infty)$ or $\rho(-\infty)$, $c(-\infty)$. Detailed calculations are given in §3 below. Only the results are described here.

Equation (1.13) gives two representations of v_f, corresponding to the choice of ϕ_+ or ϕ_-. Either can be used for the asymptotic calculation but it has been found that the results take their simplest form in the ϕ_--representation. To describe them let $R^3_+(d)$ and $R^3_-(d)$ denote the half-spaces defined by

$$R^3_\pm(d) = \{(x,y) : \pm(y - d) > 0\} \tag{1.17}$$

and let $A(+\infty)$ and $A(-\infty)$ be the acoustic propagators for the homogeneous fluids with parameters $\rho(\infty)$, $c(\infty)$ and $\rho(-\infty)$, $c(-\infty)$ respectively. Clearly

$$A(\pm\infty) = c^2(\pm\infty)\, A_0 \tag{1.18}$$

where A_0 corresponds to the special case $\rho(\infty) = 1$, $c(\infty) = 1$; i.e. $A_0 = -\Delta$.

Now fix a value of d, arbitrarily, and for each $h \in \mathcal{H}$ define a wave function $v^0_f(t,X)$ by

$$v^0_f(t,X) = \begin{cases} \exp(-i\,t\,A^{1/2}(\infty))h^+(X), & X \in R^3_+(d), \\ \exp(-i\,t\,A^{1/2}(-\infty))h^-(X), & X \in R^3_-(d), \end{cases} \tag{1.19}$$

where h^+ and h^- are the functions in $L_2(R^3)$ whose Fourier transforms

$$\hat{h}^\pm(P) = \frac{1}{(2\pi)^{3/2}} \int_{R^3} e^{-iP\cdot X}\, h^\pm(X)\,dX \tag{1.20}$$

are defined by

$$\hat{h}^+(P) = \begin{cases} c(\infty)\,\rho^{1/2}(\infty)\,\hat{h}_-(P), & P \in R^3_+(0), \\ 0, & P \in R^3_-(0), \end{cases} \tag{1.21}$$

and

$$\hat{h}^-(P) = \begin{cases} 0, & P \in R^3_+(0), \\ c(-\infty)\,\rho^{1/2}(-\infty)\,\hat{h}_-(P), & P \in R^3_-(0). \end{cases} \tag{1.22}$$

In (1.21), (1.22) the function $\hat{h}_-$ is the ϕ_--transform of $h \in \mathcal{H}$:

$$\hat{h}_-(P) = \int_{R^3} \overline{\phi_-(X,P)}\, h(X) c^{-2}(y) \rho^{-1}(y) dX. \tag{1.23}$$

With this notation the asymptotic behavior of $v_f(t,X)$ may be described by

$$\lim_{t\to+\infty} \|v_f(t,\cdot) - v_f^0(t,\cdot)\|_{\mathcal{H}} = 0. \tag{1.24}$$

Moreover, if

$$u_f^0(t,X) = \text{Re}\, \{v_f(t,X)\} \tag{1.25}$$

then (1.24) and the inequality $|\text{Re } z| \le |z|$ imply that

$$\lim_{t\to+\infty} \|u_f(t,\cdot) - u_f^0(t,\cdot)\|_{\mathcal{H}} = 0. \tag{1.26}$$

Equations (1.24) and (1.26) hold for all $h \in \mathcal{H}$. If also $A^{1/2} h \in \mathcal{H}$, so that $u_f(t,X)$ has finite energy then one also has

$$\lim_{t\to+\infty} \|D_j u_f(t,\cdot) - D_j u_f^0(t,\cdot)\|_{\mathcal{H}} = 0,\ j = 0,1,2,3. \tag{1.27}$$

Returning to the definition (1.19)-(1.23) of v_f^0, note that in the complementary half-spaces $R_+^3(d)$ and $R_-^3(d)$ it represents sound fields in homogeneous fluids with parameters $\rho(\infty)$, $c(\infty)$ and $\rho(-\infty)$, $c(-\infty)$ respectively. The latter behave for $t \to \infty$ like diverging spherical waves with speeds $c(\infty)$ and $c(-\infty)$, respectively, by the theory of [25]. Hence (1.26), (1.27) express the property that u_f behaves for large t like a diverging spherical wave.

Now consider one of the guided components $v_k(t,X)$, given by (1.14). Using the definition (3.1.23) of ψ_k the integral can be written

$$v_k(t,x,y) = \frac{1}{2\pi} \int_{\Omega_k} \exp\{i(x\cdot p - t\omega_k(|p|))\}\, \psi_k(y,|p|)\, \tilde{h}_k(p) dp. \tag{1.28}$$

For large x and t the exponential factor in the integrand is rapidly oscillating while the remaining factors are slowly varying. This suggests that stationary phase estimates can be used to approximate v_k for large t.

The phase of the exponential is a stationary function of $p \in \Omega_k \subset R^2$ if and only if

$$U_k(|p|)p/|p| = x/t \tag{1.29}$$

where

$$U_k(|p|) = \omega_k'(|p|) \tag{1.30}$$

is the group speed associated with the dispersion relation $\omega = \omega_k(|p|)$. A number of cases may arise, depending on the number of solutions p of (1.29). Here it will be assumed, for brevity, that $U_k(\mu)$ is a monotone decreasing function that maps $\mathcal{O}_k$ onto $(c_m, c(\infty))$. In this case (1.30) has a unique solution if $|x|/t$ lies in the range of $U_k(|p|)$; that is,

$$c_m < |x|/t < c(\infty), \tag{1.31}$$

and no solution otherwise. The solution is given by

$$p = Q_k(|x|/t)x/|x| \tag{1.32}$$

where $Q_k = U_k^{-1}$; that is,

$$v = U_k(\mu) \Longleftrightarrow \mu = Q_k(v). \tag{1.33}$$

The stationary phase estimates are formulated in Appendix 2. They imply that for $|x|/t$ in the interval (1.31) there is one stationary point (1.32) which makes a contribution

$$v_k^{\infty}(t,x,y,p) = \frac{|p|^{1/2} \exp\{i(|x||p| - t\omega_k(|p|))\} \psi_k(y,|p|)\tilde{h}_k(p)}{t\{U_k(|p|)\ |U_k'(|p|)|\}^{1/2}} \tag{1.34}$$

to the integral in (1.28). For $|x|/t$ outside the interval (1.31) there is no point of stationary phase. Thus the stationary phase approximation to $v_k(t,x,y)$ is

$$v_k^{\infty}(t,x,y) = \chi(|x|/t)\ v_k^{\infty}(t,x,y,Q_k(|x|/t)x/|x|) \tag{1.35}$$

where χ is the characteristic function of the interval $(c_m, c(\infty))$.

The results in Appendix 2 guarantee that if $\tilde{h}_k$ is a smooth function with compact support then $v_k^{\infty}(t,x,y)$ is a good approximation to $v_k(t,x,y)$ for large t, uniformly for all $x \in R^2 - \{0\}$ and $y \in R$. For general $h \in \mathcal{H}$ one has only $\tilde{h}_k \in L_2(\Omega_k)$ and the error estimates of Appendix 2 fail. Nevertheless, it is important to recognize that for general $h \in \mathcal{H}$ the definition (1.34), (1.35) is still meaningful and one has

(1.36) $$v_k^\infty(t,\cdot) \in \mathcal{H}$$

and

(1.37) $$\|v_k^\infty(t,\cdot)\|_{\mathcal{H}} = \|\tilde{h}_k\|_{L_2(\Omega_k)}$$

for all $t > 0$ and $1 \leq k < N_0$. These properties may be verified by direct calculation.

The central fact concerning v_k^∞ in the Hilbert space theory states that for every $h \in \mathcal{H}$ one has

(1.38) $$\lim_{t\to\infty} \|v_k(t,\cdot) - v_k^\infty(t,\cdot)\|_{\mathcal{H}} = 0.$$

It follows as in the case of v_f that if

(1.39) $$u_k^\infty(t,x,y) = \text{Re}\,\{v_k^\infty(t,x,y)\}$$

then

(1.40) $$\lim_{t\to\infty} \|u_k(t,\cdot) - u_k^\infty(t,\cdot)\|_{\mathcal{H}} = 0.$$

Moreover, if $h \in D(A^{1/2})$, so that u_k is a field with finite energy, then the derivatives $D_j\, u_k(t,X)$ are in $\mathcal{H}$ and asymptotic wave functions $u_{k,j}^\infty$ for them can be constructed such that

(1.41) $$\lim_{t\to\infty} \|D_j\, u_k(t,\cdot) - u_{k,j}^\infty(t,\cdot)\|_{\mathcal{H}} = 0$$

for $j = 0,1,2,3$. The construction of $u_{k,j}^\infty$ is given in §4 below.

Returning to the definition (1.34), (1.35) note that $v_k^\infty(t,X)$ is an inhomogeneous cylindrical wave that propagates radially and dispersively in planes $y = \text{const.}$ and is exponentially damped by the factor $\psi_k(y,|p|)$ in the y-direction. This behavior and the convergence properties (1.38), (1.41) motivate the term "guided mode" for $v_k(t,X)$.

Each sound field u with finite energy has a fixed total energy

(1.42) $$E(u,R^3,t) = E(u,R^3,0) < \infty;$$

see (2.1.9). The same is true of the corresponding modal waves:

$$(1.43)\qquad \begin{cases} E(u_f,R^3,t) = E(u_f,R^3,0), \\ E(u_k,R^3,t) = E(u_k,R^3,0). \end{cases}$$

In fact, the Parseval formula (3.1.59) will be shown to imply the energy partition formula

$$(1.44)\qquad E(u,R^3,0) = E(u_f,R^3,0) + \sum_{k=1}^{N_0-1} E(u_k,R^3,0).$$

Moreover, the asymptotic estimates (1.27), (1.41) for the modal waves u_f, u_k yield precise information on the ultimate distribution for $t \to \infty$ of the partial energies. The principal results will be formulated here without proofs. The discussion is completed in §5.

The notation

$$(1.45)\qquad E^{\infty}(u,K) = \lim_{t\to\infty} E(u,K,t)$$

will be used whenever the limit exists. A first result is the transiency of all waves with finite energy in stratified fluids:

$$(1.46)\qquad E^{\infty}(u,K) = 0 \text{ for all bounded } K \subset R^3.$$

Next consider the free component $u_f(t,X)$. (1.27) implies that in <u>each</u> half-space $R^3_+(d)$ or $R^3_-(d)$ (d is arbitrary) u_f behaves like a wave in a homogeneous fluid. Thus if C^+ (resp., C^-) denotes any cone in $R^3_+(d)$ (resp., $R^3_-(d)$) then (1.27) and the results on asymptotic wave functions of [25] imply that

$$(1.47)\qquad E^{\infty}(u_f,C^{\pm}) = c^2(\pm\infty) \int_{C^{\pm}} |P|^2 \, |\hat{h}_-(P)|^2 dP.$$

It is noteworthy that the asymptotic distribution in cones of the partial energy $E(u_f,R^3,0)$ is determined by the ϕ_--transform of the complex initial state $A^{1/2} h = A^{1/2} f + i g$. An alternative form of (1.47) is

$$(1.48)\qquad E^{\infty}(u_f,C^{\pm}) = \int_{C^{\pm}} |\omega(P)\hat{f}_-(P) + i\, \hat{g}_-(P)|^2 dP$$

where $\omega(P)$ is given by (1.15). Finally, note that one may take $C^{\pm} = R^3_{\pm}(d)$ in (1.47) and that d is arbitrary. It follows that if

(1.49) $$S = \{(x,y) : d_1 \le y \le d_2\}$$

is an arbitrary finite slab parallel to the stratification then

(1.50) $$\mathcal{E}^{\infty}(u_f,S) = 0;$$

i.e., the free component u_f ultimately propagates out of every slab (1.49).

Now consider the family of cones defined by

(1.51) $$C(\varepsilon,d) = \{(x,y) : |y - d| < \varepsilon|x|\}$$

where $d \in R$ and $\varepsilon > 0$. Then if u_k is any guided component the estimates (1.41) imply that

(1.52) $$\mathcal{E}^{\infty}(u_k,C(\varepsilon,d)) = \mathcal{E}(u_k,R^3,0)$$

for <u>every</u> d and <u>every</u> $\varepsilon > 0$. This is to be contrasted with the result (1.47) for u_f. Finally, if S is the slab defined by (1.49) then (1.41) implies that, in contrast with (1.50), one has

(1.53) $$\mathcal{E}^{\infty}(u_k,S) = \int_{\Omega_k} |\omega_k(|p|)\tilde{h}_k(p)|^2 \left(\int_{d_1}^{d_2} \psi_k^2(y,|p|)c^{-2}(y)\rho^{-1}(y)dy\right)dp.$$

Recall that the $\psi_k(y,|p|)$ are orthonormal on $-\infty < y < \infty$ with respect to $c^{-2}(y)\rho^{-1}(y)dy$. Hence the factor in parenthesis in the integrand of (1.53) lies between 0 and 1. Thus, to be sure, for example, that

(1.54) $$\mathcal{E}^{\infty}(u_k,S) \ge .9\ \mathcal{E}(u_k,R^3,0)$$

it is only necessary to choose d_1, d_2 such that

(1.55) $$\int_{d_1}^{d_2} \psi_k^2(y,|p|)c^{-2}(y)\rho^{-1}(y)dy \ge .9.$$

§2. NORMAL MODE EXPANSIONS OF TRANSIENT SOUND FIELDS

Consider a general sound field with finite energy:

(2.1) $$u(t,\cdot) = (\cos t\, A^{1/2})\, f + (A^{-1/2} \sin t\, A^{1/2})\, g$$

where $f \in D(A^{1/2})$ and $g \in \mathcal{H}$. It was shown in Chapter 3, Corollaries 9.4

and 9.10, that the orthogonal projections $\{P_+, P_-, P_0, P_1, \cdots\}$ in $\mathcal{H}$ defined by the normal modes form a complete family that reduces A. Hence the same is true of the orthogonal projections

$$P_f = P_+ + P_- + P_0 \tag{2.2}$$

and

$$P_g = \sum_{k=1}^{N_0-1} P_k. \tag{2.3}$$

This provides a decomposition

$$u(t,X) = u_f(t,X) + u_g(t,X) \tag{2.4}$$

into orthogonal partial waves

$$\begin{cases} u_f(t,\cdot) = P_f u(t,\cdot) \\ u_g(t,\cdot) = P_g u(t,\cdot). \end{cases} \tag{2.5}$$

These components are analyzed in the following two sections.

§3. TRANSIENT FREE WAVES

The normal mode expansions of §2 are used in this section to calculate the asymptotic behavior for $t \to \infty$ of the free component $u_f(t,\cdot) = P_f u(t,\cdot)$. The principal result is that in each of the half-spaces $R^3_+(d)$ and $R^3_-(d)$, $u_f(t,\cdot)$ is asymptotically equal to a wave function for a homogeneous fluid with parameters $\rho(\infty)$, $c(\infty)$ and $\rho(-\infty)$, $c(-\infty)$, respectively. It is this behavior that motivates the term "free component" for $u_f(t,\cdot)$.

It will be assumed that the total acoustic potential u satisfies $u(t,\cdot) = \text{Re}\,\{v(t,\cdot)\}$ where $v(t,\cdot) = \exp\{-i\,t\,A^{1/2}\}\,h$ and $h \in D(A^{1/2})$. The corresponding partial waves $u_k(t,\cdot) = P_k u(t,\cdot)$ with $k \geq 1$ satisfy $u_k(t,\cdot) = \text{Re}\,\{\exp(-i\,t\,A^{1/2})\,P_k h\}$. This follows from the fact that the normal mode functions $\psi_k(y,p)$ are real for $k \geq 1$, which implies that $P_k(\overline{h}) = \overline{P_k(h)}$. It follows by addition that $u_g(t,\cdot) = \text{Re}\,\{\exp(-i\,t\,A^{1/2})\,P_g h\} = \text{Re}\,\{v_g(t,\cdot)\}$ and hence

$$
(3.1) \qquad \begin{cases} u_f(t,\cdot) = \text{Re}\,\{v_f(t,\cdot)\}, \\ v_f(t,\cdot) = \exp\,(-i\,t\,A^{1/2})\,P_f h = \exp\,(-i\,t\,A^{1/2})\,h_f. \end{cases}
$$

The normal mode representations

$$
(3.2) \qquad v_f(t,x,y) = \int_{R^3} \phi_\pm(x,y,p,q)\,\exp\,\{-it\lambda^{1/2}\,(p,q)\}\,\hat{h}_\pm(p,q)dpdq
$$

provide the starting point for calculating the asymptotic behavior of $u_f(t,x,y)$ for large t.

Equation (3.2) gives two representations of v_f corresponding to the two families ϕ_+ and ϕ_-. The calculations below are based on the ϕ_--representation which has been found to yield the simplest form of the asymptotic wave function. It will be convenient to introduce the characteristic functions χ_+, χ_0 and χ_- of the cones C_+, C_0 and C_- in (p,q)-space and to decompose $\hat{h}_-$ as

$$
(3.3) \qquad \hat{h}_-(p,q) = \ell(p,q) + m(p,q) + n(p,q)
$$

where $\ell = \chi_+\,\hat{h}_-$, $m = \chi_0\,\hat{h}_-$ and $n = \chi_-\,\hat{h}_-$. The corresponding decomposition of v_f is

$$
(3.4) \qquad v_f = v_\ell + v_m + v_n
$$

where

$$
(3.5) \qquad \begin{cases} v_\ell = \exp\,(-i\,t\,A^{1/2})\,\Phi_-^*\ell \\ v_m = \exp\,(-i\,t\,A^{1/2})\,\Phi_-^*m \\ v_n = \exp\,(-i\,t\,A^{1/2})\,\Phi_-^*n. \end{cases}
$$

The behavior for $t \to \infty$ of these three functions will be analyzed separately.

<u>Behavior of v_ℓ</u>. The partial wave v_ℓ has the representation

$$
(3.6) \qquad v_\ell(t,x,y) = \int_{C_+} \phi_-(x,y,p,q)\,\exp\,(-i\,t\,\omega_+(p,q))\,\ell(p,q)dpdq
$$

where

$$\omega_{\pm}(p,q) = c(\pm\infty)\sqrt{|p|^2+q^2}. \tag{3.7}$$

(Recall $\lambda(p,q) = \omega_{\pm}^2(p,q)$ for $\pm q > 0$.) To discover the behavior of $v_{\ell}(t,x,y)$ for $(x,y) \in R_{\pm}^2(d)$ and $t \to \infty$ it will be convenient to write $\phi_{-}(x,y,p,q)$ in a way that puts in evidence its behavior for $y \to \pm\infty$. To this end recall that by (3.1.38)-(3.1.45), one has

$$\phi_{-}(x,y,p,q) = (2\pi)^{-1}\, c(\infty)(2q)^{1/2}\, e^{ip\cdot x}\, \overline{\psi_{+}(y,|p|,\lambda)} \tag{3.8}$$

for $(p,q) = X_{+}(p,\lambda) \in C_{+}$ (and hence $\lambda = \lambda(p,q) = c^2(\infty)(|p|^2+q^2)$). Moreover, by (3.5.5), (3.5.6) and (3.9.41) one can write

$$\begin{aligned}\psi_{+}(y,\mu,\lambda) &= \left(\frac{\rho(\infty)}{4\pi q_{+}(\mu,\lambda)}\right)^{1/2} T_{+}(\mu,\lambda)\, \phi_4(y,\mu,\lambda) \\ &= \left(\frac{\rho(\infty)}{4\pi q_{+}(\mu,\lambda)}\right)^{1/2} T_{+}(y,\mu,\lambda)\, \exp\{-i\,y\,q_{-}(\mu,\lambda)\}\end{aligned} \tag{3.9}$$

where

$$T_{+}(y,\mu,\lambda) = T_{+}(\mu,\lambda)\, \phi_4(y,\mu,\lambda)\, \exp\{i\,y\,q_{-}(\mu,\lambda)\} \to T_{+}(\mu,\lambda),\ y \to -\infty. \tag{3.10}$$

Similarly, by (3.5.1) and (3.5.6)-(3.5.11) one can write

$$\begin{aligned}\psi_{+}(y,\mu,\lambda) = \left(\frac{\rho(\infty)}{4\pi q_{+}(\mu,\lambda)}\right)^{1/2} &[I_{+}(y,\mu,\lambda)\, \exp\{-i\,y\,q_{+}(\mu,\lambda)\} \\ &+ R_{+}(y,\mu,\lambda)\, \exp\{i\,y\,q_{+}(\mu,\lambda)\}]\end{aligned} \tag{3.11}$$

where

$$\left.\begin{aligned} I_{+}(y,\mu,\lambda) &= \phi_2(y,\mu,\lambda)\, \exp\{i\,y\,q_{+}(\mu,\lambda)\} \to 1 \\ R_{+}(y,\mu,\lambda) &= R_{+}(\mu,\lambda)\, \phi_1(y,\mu,\lambda)\, \exp\{-i\,y\,q_{+}(\mu,\lambda)\} \to R_{+}(\mu,\lambda)\end{aligned}\right\}\ y \to +\infty. \tag{3.12}$$

Combining (3.6), (3.8) and (3.11) gives

$$\begin{aligned} v_{\ell}(t,x,y) \\ = c(\infty)\rho(\infty)^{1/2}(2\pi)^{-3/2} \int_{C_{+}} \exp\{i(x\cdot p + y\,q - t\omega_{+}(p,q))\}\, \overline{I_{+}(y,|p|,\lambda)}\, \ell(p,q)\,dp\,dq\end{aligned} \tag{3.13}$$

$$+ c(\infty)\rho(\infty)^{1/2} (2\pi)^{-3/2} \int_{C_+} \exp\{i(x\cdot p - yq - t\omega_+(p,q)\} \overline{R_+(y,|p|,\lambda)}\, \ell(p,q)dpdq.$$

(3.13 cont.)

It is natural to expect that in $R^3_+(d)$ the partial wave $v_\ell(t,x,y)$ will propagate as $t \to \infty$ into regions where y is large and hence $I_+(y,|p|,\lambda)$ and $R_+(y,|p|,\lambda)$ are near their limiting values. Thus the representation (3.13) suggests the conjecture that

$$v_\ell(t,\cdot) \sim v^0_\ell(t,\cdot) + v^1_\ell(t,\cdot) \text{ in } L_2(R^3_+(d)),\ t \to \infty \tag{3.14}$$

where v^0_ℓ and v^1_ℓ are defined by

$$v^0_\ell(t,x,y) = c(\infty)\rho(\infty)^{1/2} (2\pi)^{-3/2} \int_{C_+} \exp\{i(x\cdot p + yq - t\omega_+(p,q)\}\ell(p,q)dpdq \tag{3.15}$$

and

$$\begin{aligned} v^1_\ell(t,x,y) &= c(\infty)\rho(\infty)^{1/2} (2\pi)^{-3/2} \int_{C_+} \exp\{i(x\cdot p - yq - t\omega_+(p,q)\} \times \\ &\qquad \times \overline{R_+(|p|,\lambda)}\, \ell(p,q)dpdq \\ &= c(\infty)\rho(\infty)^{1/2} (2\pi)^{-3/2} \int_{-C_+} \exp\{i(x\cdot p + yq - t\omega_+(p,q)\} \times \\ &\qquad \times \overline{R_+(|p|,\lambda)}\, \ell(p,-q)dpdq \end{aligned} \tag{3.16}$$

where $-C_+ = \{(p,q) : (p,-q) \in C_+\} = \{(p,q) : q < -a|p|\}$. Note that v^0_ℓ and v^1_ℓ are waves in a homogeneous medium with density $\rho(\infty)$ and sound speed $c(\infty)$. More precisely,

$$\begin{cases} v^0_\ell(t,\cdot) = \exp(-i\,t\,c(\infty)\,A_0^{1/2})\,h_\ell \\ v^1_\ell(t,\cdot) = \exp(-i\,t\,c(\infty)\,A_0^{1/2})\,h^1_\ell \end{cases} \tag{3.17}$$

where A_0 is the selfadjoint realization in $L_2(R^3)$ of $-\Delta = -(\partial^2/\partial x_1^2 + \partial^2/\partial x_2^2 + \partial^2/\partial y^2)$ and h_ℓ and h^1_ℓ are the functions in $L_2(R^3)$ whose Fourier transforms are

$$(3.18)\quad \begin{cases} \hat{h}_\ell(p,q) = c(\infty)\rho(\infty)^{1/2}\, \ell(p,q) = c(\infty)\rho(\infty)^{1/2}\, \chi_+(p,q)\, \hat{h}_-(p,q), \\ \hat{h}^1_\ell(p,q) = c(\infty)\rho(\infty)^{1/2}\, \overline{R_+(|p|,\lambda)}\, \ell(p,-q) \\ \qquad = c(\infty)\rho(\infty)^{1/2}\, \overline{R_+(|p|,\lambda)}\, (1 - \chi_+(p,q))\, \hat{h}_-(p,-q). \end{cases}$$

Both functions are in $L_2(R^3)$ because $\hat{h}_- \in L_2(R^3)$ and $|R_+(|p|,\lambda)| \le \rho^{1/2}(\infty)$ by the conservation law (3.1.31). Moreover, supp $\hat{h}^1_\ell \subset -C_+$ and hence the theory of asymptotic wave functions for d'Alembert's equation [25, Ch.2] implies that $v^1_\ell(t,\cdot) \sim 0$ in $L_2(R^3_+(d))$ when $t \to \infty$. Combining this with (3.14) gives

$$(3.19)\qquad v_\ell(t,\cdot) \sim v^0_\ell(t,\cdot) \text{ in } L_2(R^3_+(d)),\ t \to \infty.$$

Now consider the behavior of $v_\ell(t,x,y)$ for $(x,y) \in R^3_-(d)$, $t \to \infty$. Combining (3.6), (3.9) and (3.10) gives

$$v_\ell(t,x,y)$$
$$(3.20)$$
$$= c(\infty)\, \rho(\infty)^{1/2}\, (2\pi)^{-3/2} \int_{C_+} \exp\{i(x\cdot p + y q_- - t\omega_+)\}\, \overline{T_+(y,|p|,\lambda)}\, \ell(p,q)\,dpdq$$

where $q_- = q_-(|p|,\lambda)$, $\omega_+ = \omega_+(p,q)$ and $\lambda = \lambda(p,q) = \omega^2_+(p,q)$. This representation suggests that

$$(3.21)\qquad v_\ell(t,\cdot) \sim v^2_\ell(t,\cdot) \text{ in } L_2(R^3_-(d)),\ t \to \infty$$

where

$$v^2_\ell(t,x,y)$$
$$(3.22)$$
$$= c(\infty)\, \rho(\infty)^{1/2}\, (2\pi)^{-3/2} \int_{C_+} \exp\{i(x\cdot p + y q_- - t\omega_+)\}\, \overline{T_+(|p|,\lambda)}\, \ell(p,q)\,dpdq.$$

Now the mapping $(p,q) \to (p,q') = X'(p,q) = (p, q_-(|p|,\omega^2_+(p,q)))$ with domain C_+ has range $X'(C_+) = R^3_+ = R^3_+(0)$, Jacobian $\partial(p,q)/\partial(p,q') = c^2(-\infty)q'/c^2(\infty)q$ and satisfies $\omega_+(p,q) = \omega_-(p,q')$. Thus (3.22) implies the representation

$$v_\ell^2(t,x,y) = c(\infty)\rho(\infty)^{1/2}(2\pi)^{-3/2}\int_{R_+^3} \exp\{i(x\cdot p + y q' - t\omega_-)\} \times$$

(3.23)

$$\times\ \overline{T_+(|p|,\omega_-^2(p,q'))}\ \ell(p,q)(c^2(-\infty)q'/c^2(\infty)q)\,dp\,dq'$$

where $q = q(|p|,q') = \sqrt{a^2(|p|^2 + q'^2) + q'^2}$. Note that

$$v_\ell^2(t,\cdot) = \exp(-i\,t\,c(-\infty)\,A_0^{1/2})\,h_\ell^2 \tag{3.24}$$

where $h_\ell^2 \in L_2(R^3)$ has Fourier transform

$$\hat{h}_\ell^2(p,q') = c(\infty)\rho(\infty)^{1/2}\ \overline{T_+(|p|,\omega_-^2(p,q'))}\ \ell(p,q)(c^2(-\infty)q'/c^2(\infty)q). \tag{3.25}$$

Since supp $\hat{h}_\ell^2 \subset R_+^3$ the results of [25, Ch. 2] imply that $v_\ell^2(t,\cdot) \sim 0$ in $L_2(R_-^3(d))$ when $t \to \infty$. Combining this with (3.21) gives

$$v_\ell(t,\cdot) \sim 0 \text{ in } L_2(R_-^3(d)),\ t \to \infty. \tag{3.26}$$

Analogous conjectures concerning $v_m(t,\cdot)$ and $v_n(t,\cdot)$ will now be formulated. Only the main steps of the calculations will be given since the method is the same as for $v_\ell(t,\cdot)$.

<u>Behavior of v_m</u>. v_m has the representation

$$v_m(t,x,y) = \int_{C_0} \phi_-(x,y,p,q)\exp(-i\,t\,\omega_+(p,q))\,m(p,q)\,dp\,dq, \tag{3.27}$$

by (3.5), where

$$\phi_-(x,y,p,q) = (2\pi)^{-1}c(\infty)(2q)^{1/2}e^{ip\cdot x}\ \overline{\psi_0(y,|p|,\lambda)} \tag{3.28}$$

for $(p,q) = X_0(p,\lambda) \in C_0$ (and hence $\lambda = \lambda(p,q) = \omega_+^2(p,q)$). Moreover (see (3.5.18)-(3.5.24))

$$\psi_0(y,\mu,\lambda) = \left(\frac{\rho(\infty)}{4\pi q_+(\mu,\lambda)}\right)^{1/2} T_0(y,\mu,\lambda)\exp(y\,q_-'(\mu,\lambda)) \tag{3.29}$$

where

$$T_0(y,\mu,\lambda) = T_0(\mu,\lambda)\,\phi_3(y,\mu,\lambda)\exp(-y\,q_-'(\mu,\lambda)) \to T_0(\mu,\lambda),\ y \to -\infty \tag{3.30}$$

and

$$\psi_0(y,\mu,\lambda) = \left(\frac{\rho(\infty)}{4\pi q_+(\mu,\lambda)}\right)^{1/2} [I_0(y,\mu,\lambda) \exp\{-i y q_+(\mu,\lambda)\} \tag{3.31}$$

$$+ R_0(y,\mu,\lambda) \exp\{i y q_+(\mu,\lambda)\}]$$

where

$$\left.\begin{cases} I_0(y,\mu,\lambda) = \phi_2(y,\mu,\lambda) \exp\{i y q_+(\mu,\lambda)\} \to 1 \\ R_0(y,\mu,\lambda) = R_0(\mu,\lambda)\, \phi_1(y,\mu,\lambda) \exp\{-i y q_+(\mu,\lambda)\} \to R_0(\mu,\lambda) \end{cases}\right\} y \to \infty. \tag{3.32}$$

Combining (3.27), (3.28) and (3.29) gives

$$v_m(t,x,y) = c(\infty)\rho(\infty)^{1/2} (2\pi)^{-3/2} \int_{C_0} \exp\{i(x \cdot p - t\omega_+)\} \times \tag{3.33}$$

$$\times \overline{T_0(y,|p|,\lambda)} \exp(y\, q'_-)\, m(p,q)\,dp\,dq$$

where $\omega_+ = \omega_+(p,q)$, $\lambda = \omega_+^2(p,q)$ and $q'_- = q'_-(|p|,\lambda)$. Since

$$T_0(y,|p|,\lambda) \exp(y\, q'_-(|p|,\lambda)) \to 0, \quad y \to -\infty, \tag{3.34}$$

equation (3.33) suggests that

$$v_m(t,\cdot) \sim 0 \text{ in } L_2(R^3_-(d)), \quad t \to \infty. \tag{3.35}$$

Similarly, combining (3.27), (3.28) and (3.31) gives

$$\begin{aligned} v_m(t,x,y) = {} & c(\infty)\rho(\infty)^{1/2} (2\pi)^{-3/2} \int_{C_0} \exp\{i(x \cdot p + y q - t\omega_+)\} \times \\ & \times \overline{I_0(y,|p|,\lambda)}\, m(p,q)\,dp\,dq \\ & + c(\infty)\rho(\infty)^{1/2} (2\pi)^{-3/2} \int_{C_0} \exp\{i(x \cdot p - y q - t\omega_+)\} \times \\ & \times \overline{R_0(y,|p|,\lambda)}\, m(p,q)\,dp\,dq \end{aligned} \tag{3.36}$$

which suggests that

$$v_m(t,\cdot) \sim v_m^0(t,\cdot) + v_m^1(t,\cdot) \text{ in } L_2(R^3_+(d)), \quad t \to \infty \tag{3.37}$$

where v_m^0 and v_m^1 are defined by

$$\text{(3.38)} \qquad \begin{cases} v_m^0(t,\cdot) = \exp\,(-i\,t\,c(\infty)\,A_0^{1/2})\,h_m \\ v_m^1(t,\cdot) = \exp\,(-i\,t\,c(\infty)\,A_0^{1/2})\,h_m^1 \end{cases}$$

and h_m and h_m^1 are the functions in $L_2(R^3)$ whose Fourier transforms are

$$\text{(3.39)} \qquad \begin{cases} \hat{h}_m(p,q) = c(\infty)\rho(\infty)^{1/2}\, m(p,q) = c(\infty)\rho(\infty)^{1/2}\,\chi_0(p,q)\,\hat{h}_-(p,q), \\ \hat{h}_m^1(p,q) = c(\infty)\rho(\infty)^{1/2}\,\overline{R_0(|p|,\lambda)}\, m(p,-q) \end{cases}$$

$$= c(\infty)\rho(\infty)^{1/2}\,\overline{R_0(|p|,\lambda)}\,(1 - \chi_0(p,q))\,\hat{h}_-(p,-q).$$

Note that $\operatorname{supp} \hat{h}_m^1 \subset R_-^3$ and hence $v_m^1(t,\cdot) \sim 0$ in $L_2(R_+^3(d))$ when $t \to \infty$. Combining this with (3.37) gives

$$\text{(3.40)} \qquad v_m(t,\cdot) \sim v_m^0(t,\cdot) \text{ in } L_2(R_+^3(d)),\ t \to \infty.$$

<u>Behavior of v_n.</u> v_n has the representation

$$\text{(3.41)} \qquad v_n(t,x,y) = \int_{C_-} \phi_-(x,y,p,q)\,\exp\,(-i\,t\,\omega_-(p,q))\,n(p,q)dpdq,$$

by (3.5), where

$$\text{(3.42)} \qquad \phi_-(x,y,p,q) = (2\pi)^{-1}\, c(-\infty)(2|q|)^{1/2}\, e^{ip\cdot x}\,\overline{\psi_-(y,|p|,\lambda)}$$

for $(p,q) = X_-(p,\lambda) \in C_-$ (and hence $\lambda = \lambda(p,q) = \omega_-^2(p,q)$). Moreover (see (3.5.5) and (3.5.12)-(3.5.17)),

$$\text{(3.43)} \qquad \psi_-(y,\mu,\lambda) = \left(\frac{\rho(-\infty)}{4\pi q_-(\mu,\lambda)}\right)^{1/2} T_-(y,\mu,\lambda)\,\exp\,\{i\,y\,q_+(\mu,\lambda)\}$$

where

$$\text{(3.44)} \qquad T_-(y,\mu,\lambda) = T_-(\mu,\lambda)\,\phi_1(y,\mu,\lambda)\,\exp\,\{-i\,y\,q_+(\mu,\lambda)\} \to T_-(\mu,\lambda),\ y \to \infty$$

and

$$\psi_-(y,\mu,\lambda) = \left(\frac{\rho(-\infty)}{4\pi q_-(\mu,\lambda)}\right)^{1/2} [I_-(y,\mu,\lambda) \exp\{i y q_-(\mu,\lambda)\} + R_-(y,\mu,\lambda) \exp\{-i y q_-(\mu,\lambda)\}] \tag{3.45}$$

where

$$\left.\begin{cases} I_-(y,\mu,\lambda) = \phi_3(y,\mu,\lambda) \exp\{-i y q_-(\mu,\lambda)\} \to 1 \\ R_-(y,\mu,\lambda) = R_-(\mu,\lambda)\, \phi_4(y,\mu,\lambda) \exp\{i y q_-(\mu,\lambda)\} \to R_-(\mu,\lambda) \end{cases}\right\} y \to -\infty. \tag{3.46}$$

Combining (3.41), (3.42) and (3.45) gives, after simplification using $q_-(|p|,\omega_-^2(p,q)) = (q^2)^{1/2} = -q$ for $(p,q) \in C_-$,

$$\begin{aligned} v_n(t,x,y) &= c(-\infty)\rho(-\infty)^{1/2} (2\pi)^{-3/2} \int_{C_-} \exp\{i(x\cdot p + yq - t\omega_-)\} \times \\ &\quad \times \overline{I_-(y,|p|,\lambda)}\, n(p,q)\,dpdq \\ &+ c(-\infty)\rho(-\infty)^{1/2} (2\pi)^{-3/2} \int_{C_-} \exp\{i(x\cdot p - yq - t\omega_-)\} \times \\ &\quad \times \overline{R_-(y,|p|,\lambda)}\, n(p,q)\,dpdq. \end{aligned} \tag{3.47}$$

This suggests the asymptotic behavior

$$v_n(t,\cdot) \sim v_n^0(t,\cdot) + v_n^1(t,\cdot) \text{ in } L_2(R_-^3(d)),\ t \to \infty, \tag{3.48}$$

where

$$\begin{cases} v_n^0(t,\cdot) = \exp(-i t c(-\infty) A_0^{1/2})\, h_n \\ v_n^1(t,\cdot) = \exp(-i t c(-\infty) A_0^{1/2})\, h_n^1 \end{cases} \tag{3.49}$$

and h_n and h_n^1 are the functions whose Fourier transforms are

$$\begin{cases} \hat{h}_n(p,q) = c(-\infty)\rho(-\infty)^{1/2} n(p,q) = c(-\infty)\rho(-\infty)^{1/2} \chi_-(p,q)\hat{h}_-(p,q), \\ \hat{h}_n^1(p,q) = c(-\infty)\rho(-\infty)^{1/2}\, \overline{R_-(|p|,\lambda)}\, n(p,-q) \\ \qquad = c(-\infty)\rho(-\infty)^{1/2}\, \overline{R_-(|p|,\lambda)}\, (1 - \chi_-(p,q))\, \hat{h}_-(p,-q). \end{cases} \tag{3.50}$$

Note that supp $\hat{h}_n^1 \subset R_+^3$ and hence $v_n^1(t,\cdot) \sim 0$ in $L_2(R_-^3(d))$ when $t \to \infty$. Combining this with (3.48) gives

$$v_n(t,\cdot) \sim v_n^0(t,\cdot) \text{ in } L_2(R_-^3(d)),\ t \to \infty. \tag{3.51}$$

Finally, combining (3.41), (3.42) and (3.43) gives

$$v_n(t,x,y) = c(-\infty)\rho(-\infty)^{1/2}(2\pi)^{-3/2}\int_{C_-} \exp\{i(x\cdot p - y q_+ - t\omega_-)\} \times \overline{T_-(y,|p|,\lambda)}\, n(p,q)\,dp\,dq \tag{3.52}$$

where $\omega_- = \omega_-(p,q)$, $\lambda = \omega_-^2(p,q)$ and $q_+ = q_+(|p|,\lambda)$. This suggests that

$$v_n(t,\cdot) \sim v_n^2(t,\cdot) \text{ in } L_2(R_+^3(d)),\ t \to \infty \tag{3.53}$$

where

$$v_n^2(t,x,y) = c(-\infty)\rho(-\infty)^{1/2}(2\pi)^{-3/2}\int_{C_-} \exp\{i(x\cdot p - y q_+ - t\omega_-)\} \times \overline{T_-(|p|,\lambda)}\, n(p,q)\,dp\,dq. \tag{3.54}$$

Now the mapping $(p,q) \to (p,q') = X''(p,q) = (p,-q_+(p,\lambda(p,q)))$ maps C_- onto $X''(C_-) = -C_+$, has Jacobian $\partial(p,q)/\partial(p,q') = c^2(\infty)q'/c^2(-\infty)q$ and satisfies $\omega_-(p,q) = \omega_+(p,q')$. Thus

$$v_n^2(t,\cdot) = \exp(-i\,t\,c(\infty)\,A_0^{1/2})\,h_n^2 \tag{3.55}$$

where h_n^2 has the Fourier transform

$$\hat{h}_n^2(p,q') = c(-\infty)\rho(-\infty)^{1/2}\,\overline{T_-(|p|,\omega_+^2(p,q'))}\, n(p,q(p,q'))(c^2(\infty)q'/c^2(-\infty)q). \tag{3.56}$$

Moreover, supp $\hat{h}_n^2 \subset R_-^3$ and hence $v_n^2(t,\cdot) \sim 0$ in $L_2(R_+^3(d))$, $t \to \infty$. Combining this with (3.53) gives

$$v_n(t,\cdot) \sim 0 \text{ in } L_2(R_+^3(d)),\ t \to \infty. \tag{3.57}$$

The asymptotic behavior of $v_f(t,\cdot)$ for $t \to \infty$ may be obtained from the three cases analyzed above by superposition, equation (3.4). Thus equations (3.19), (3.26), (3.35), (3.40), (3.51) and (3.57) imply

$$(3.58)\qquad v_f(t,\cdot) \sim \begin{cases} v_\ell^0(t,\cdot) + v_m^0(t,\cdot) & \text{in } L_2(R_+^3(d)) \\ v_n^0(t,\cdot) & \text{in } L_2(R_-^3(d)) \end{cases} \quad t \to \infty.$$

On combining this with the definitions of v_ℓ^0, v_m^0 and v_n^0, equations (3.17), (3.18), (3.38), (3.39), (3.49) and (3.50), one is led to formulate

Theorem 3.1. For every $h \in \mathcal{H}$ let $v_f^0(t,\cdot)$ be defined by

$$(3.59)\qquad v_f^0(t,x,y) = \begin{cases} \exp\,(-i\,t\,c(\infty)\,A_0^{1/2})\,h^+(x,y), & (x,y) \in R_+^3(d), \\ \exp\,(-i\,t\,c(-\infty)\,A_0^{1/2})\,h^-(x,y), & (x,y) \in R_-^3(d), \end{cases}$$

where h^+ and h^- are the functions in $L_2(R^3)$ whose Fourier transforms are given by

$$(3.60)\qquad \hat{h}^+(p,q) = \begin{cases} c(\infty)\rho(\infty)^{1/2}\,\hat{h}_-(p,q), & (p,q) \in R_+^3, \\ 0, & (p,q) \in R_-^3, \end{cases}$$

and

$$(3.61)\qquad \hat{h}^-(p,q) = \begin{cases} 0, & (p,q) \in R_+^3, \\ c(-\infty)\rho(-\infty)^{1/2}\,\hat{h}_-(p,q), & (p,q) \in R_-^3. \end{cases}$$

Then

$$(3.62)\qquad \lim_{t\to\infty} \|v_f(t,\cdot) - v_f^0(t,\cdot)\|_{\mathcal{H}} = 0.$$

Theorem 3.1 implies corresponding asymptotic estimates for the free component $u_f(t,\cdot) = P_f u(t,\cdot) = \text{Re}\,\{v_f(t,\cdot)\}$ of the acoustic potential $u(t,\cdot)$. Indeed, if $u_f^0(t,\cdot)$ is defined by

$$(3.63)\qquad u_f^0(t,\cdot) = \text{Re}\,\{v_f(t,\cdot)\}$$

then Theorem 3.1 and the elementary inequality $|\text{Re}\ z| \le |z|$ imply

Corollary 3.2. For all $h \in \mathcal{H}$ one has

$$\lim_{t\to\infty} \|u_f(t,\cdot) - u_f^0(t,\cdot)\|_{\mathcal{H}} = 0. \tag{3.64}$$

If the initial state h has derivatives in $\mathcal{H}$ then $u_f(t,\cdot)$ and $u_f^0(t,\cdot)$ have the same derivatives in $\mathcal{H}$ and (3.64) can be strengthened to include these derivatives. In particular, one has

Corollary 3.3. For all $h \in L_2^1(R^3) = D(A^{1/2})$ one has

$$\lim_{t\to\infty} \|D_j u_f(t,\cdot) - D_j u_f^0(t,\cdot)\|_{\mathcal{H}} = 0, \quad j = 0,1,2,3. \tag{3.65}$$

Equation (3.65) is equivalent to convergence in energy:

$$\lim_{t\to\infty} E(u_f - u_f^0, R^3, t) = 0. \tag{3.66}$$

Corollary 3.3 can be proved by applying the method of this section to the derivatives $D_j\, v_f(t,\cdot)$ ($j = 0,1,2,3$) which are given by integrals of the same form as (3.2). Detailed proofs for the case of the Pekeris profile were given in [27].

Proof of Theorem 3.1. The remainder of this section is devoted to the proof of Theorem 3.1. The decomposition (3.3) is used for the proof. Moreover, for brevity, only the asymptotic equality (3.53) for $v_n(t,\cdot)$ is proved. The remaining five cases, namely (3.14), (3.21), (3.35), (3.40) and (3.48) can be proved by the method used for (3.53). As a first step, (3.53) will be proved for the special case of $n(p,q) \in C_0(C_-)$, the set of continuous function with compact supports in the open cone C_-. The general case will then be proved by using the fact that $C_0(C_-)$ is dense in $L_2(C_-)$.

For functions $n(p,q) \in C_0(C_-)$ the integrals defining v_n and v_n^2 converge point-wise, as well as in $\mathcal{H}$, and one can write

$$v_n(t,x,y) - v_n^2(t,x,y) = c(-\infty)\rho(-\infty)^{1/2}\,(2\pi)^{-3/2} \int_{R^2} \exp\,(ix\cdot p)w(y,p,t)dp \tag{3.67}$$

where

$$w(y,p,t) = \int_{-\infty}^{0} \exp\,\{-i(y\,q_+ + t\omega_-)\}[\overline{T_-(y,|p|,\omega_-^2)} - \overline{T_-(|p|,\omega_-^2)}]n(p,q)dq. \tag{3.68}$$

Parseval's formula in $L_2(R^2)$, applied to (3.67), gives

$$\int_{R^2} |v_n(t,x,y) - v_n^2(t,x,y)|^2\, dx = c^2(-\infty)\rho(-\infty)(2\pi)^{-1} \int_{R^2} |w(y,p,t)|^2\, dp. \tag{3.69}$$

On integrating this over $y \geq d$ one finds

$$(3.70) \quad \|v_n(t,\cdot) - v_n^2(t,\cdot)\|_{L_2(R_+^3(d))} = c(-\infty)\rho(-\infty)^{1/2}(2\pi)^{-1/2}\|w(\cdot,t)\|_{L_2(R_+^3(d))}.$$

The last relation implies that to prove (3.53) it is sufficient to prove that

$$(3.71) \quad w(\cdot,t) \to 0 \text{ in } L_2(R_+^3(d)),\ t \to \infty.$$

To this end it will be convenient to change the variable of integration in (3.68) from q to $\omega = \omega_-(p,q) = c(-\infty)\sqrt{|p|^2 + q^2}$. Solving this equation for $q < 0$ gives $q = -(\omega^2 c^{-2}(-\infty) - |p|^2)^{1/2} = -q_-(|p|,\omega^2)$ with $\omega > c(-\infty)|p|$. Hence (3.68) can be written

$$(3.72) \quad w(y,p,t) = \int_{c(-\infty)|p|}^{\infty} \exp(-i t \omega)\, W(y,p,\omega)\, d\omega$$

where

$$(3.73) \quad W(y,p,\omega) = \frac{c^{-2}(-\infty)\exp\{-i y q_+(|p|,\omega^2)\}[\overline{T_-(y,|p|,\omega^2)} - \overline{T_-(|p|,\omega^2)}]n(p,-q_-(|p|,\omega^2))\omega}{q_-(|p|,\omega^2)}.$$

The assumption that $n \in C_0(C_-)$, together with (3.44), implies that $W \in C_0(R \times \Gamma)$ where $\Gamma = \{(p,\omega) : \omega > c(-\infty)|p|\}$. Moreover, by a standard partition of unity argument, one may assume without loss of generality that

$$(3.74) \quad \operatorname{supp} W(y,\cdot) \subset \{(p,\omega) : |p| \leq p_0 \text{ and } 0 < \omega_0 \leq \omega \leq \omega_1\}$$

for all $y \in R$ where $\omega_0 > c(-\infty)p_0$. This in turn implies that

$$(3.75) \quad w(y,p,t) = \int_{\omega_0}^{\omega_1} \exp(-i t \omega)\, W(y,p,\omega)\, d\omega$$

and

$$(3.76) \quad \operatorname{supp} w(y,\cdot,t) \subset B(p_0) = \{p : |p| \leq p_0\}$$

for all $y \in R$ and $t \in R$. Thus

$$\|w(\cdot,t)\|^2_{L_2(R^3_+(d))} = \int_d^\infty \int_{B(p_0)} |w(y,p,t)|^2 \, dpdy$$

(3.77)

$$= \int_d^{y_0} \int_{B(p_0)} |w(y,p,t)|^2 \, dpdy + \int_{y_0}^\infty \int_{B(p_0)} |w(y,p,t)|^2 \, dpdy$$

for any $y_0 > d$. The proof of (3.71) will be derived from (3.77) and the following two lemmas.

Lemma 3.4. Let $n \in C_0(C_-)$ and assume that (3.74) holds. Then for each $d \in R$, $y_0 > d$ and $p_0 > 0$ one has

(3.78)
$$\lim_{t\to\infty} w(y,p,t) = 0,$$

uniformly for all $(y,p) \in [d,y_0] \times B(p_0)$.

Lemma 3.5. Under the hypotheses of Lemma 3.4 there is a constant $C = C(n)$ such that

(3.79)
$$|w(y,p,t)| \le C\, y^{1-\alpha}$$

for all $y > 0$, $p \in B(p_0)$ and $t \in R$, where $\alpha > 3/2$ is the constant of condition (1.1).

Proof of Lemma 3.4. The proof is based on a well-known proof of the Riemann-Lebesgue lemma. Note that by (3.75) one has

$$w(y,p,t) = -\int_{\omega_0-(\pi/t)}^{\omega_1-(\pi/t)} \exp(-i\omega t)\, W(y,p,\omega+(\pi/t))\,d\omega$$

(3.80)

$$= \frac{1}{2}\int_{\omega_0}^{\omega_1} \exp(-i t\omega)[W(y,p,\omega) - W(y,p,\omega+(\pi/t))]\,d\omega$$

$$- \frac{1}{2}\int_{\omega_0-(\pi/t)}^{\omega_0} \exp(-i t\omega)\, W(y,p,\omega+(\pi/t))\,d\omega$$

$$+ \frac{1}{2}\int_{\omega_1-(\pi/t)}^{\omega_1} \exp(-i t\omega)\, W(y,p,\omega+(\pi/t))\,d\omega.$$

The limit relation (3.78) is obvious from (3.80) and the continuity of W. The uniformity of the limit follows from (3.80) and the uniform continuity of W on compact subsets of $R \times \Gamma$.

Proof of Lemma 3.5. Note that by (3.72), (3.73) and (3.44) one has the estimate

$$|w(y,p,t)| \le \int_{\omega_0}^{\omega_1} |\phi_1(y,|p|,\omega^2) \exp\{-i y q_+(|p|,\omega^2)\} - 1| \, M(p,\omega) d\omega \tag{3.81}$$

for all y, p and t where

$$M(p,\omega) = c^{-2}(-\infty) \, |T_-(|p|,\omega^2)| \left| \frac{n(p,-q_-(|p|,\omega^2))\omega}{q_-(|p|,\omega^2)} \right| \tag{3.82}$$

is continuous for $|p| \le p_0$, $\omega_0 \le \omega \le \omega_1$. It follows that for any $y \in R$, $p \in B(p_0)$ and $t \in R$

$$|w(y,p,t)| \le M_0 \sup |\phi_1(y,|p|,\omega^2) \exp\{-i y q_+(|p|,\omega^2)\} - 1| \tag{3.83}$$

where $M_0 = M_0(n) = \sup M(p,\omega)$ and the suprema are taken over all $|p| \le p_0$ and $\omega_0 \le \omega \le \omega_1$. The proof of (3.79) will be based on (3.83), conditions (2.1.4) and (1.1) on $\rho(y)$ and $c(y)$ and the proof in Chapter 3 of Theorem 3.1. Note that because of the continuity of w it will suffice to prove (3.79) for all $y > y_1$ where $y_1 = y_1(n)$ is a positive constant.

The solution $\phi_1(y,\mu,\lambda)$ with $\lambda > c^2(-\infty)\mu^2 \ge c^2(\infty)\mu^2$ satisfies (see (3.3.44)ff.)

$$\phi_1(y,\mu,\lambda) = \exp\{i y q_+(\mu,\lambda)\} (\eta_1 + \eta_2) \tag{3.84}$$

where $\eta = (\eta_1,\eta_2)$ is characterized on $y \ge y_1$ as the unique solution of the integral equation (3.3.64) which can be written

$$\eta = \eta^0 + K(\mu,\lambda)\eta, \quad \eta^0 = (1,0). \tag{3.85}$$

The kernel $K(y,y',\mu,\lambda) = (K_{ij}(y,y',\mu,\lambda))$ is defined by

$$K_{1j}(y,y',\mu,\lambda) = \begin{cases} 0, & y_1 \le y' < y, \\ -E_{1j}(y',\mu,\lambda), & y' \ge y, \end{cases} \tag{3.86}$$

and

$$(3.87)\quad K_{2j}(y,y',\mu,\lambda) = \begin{cases} 0, & y_1 \le y' < y, \\ -\exp\{-2i(y-y')q_+(\mu,\lambda)\} E_{2j}(y',\mu,\lambda), & y' \ge y, \end{cases}$$

where ((3.3.43))

$$(3.88)\qquad E(y,\mu,\lambda) = B^{-1}(\mu,\lambda)\, N(y,\mu,\lambda)\, B(\mu,\lambda).$$

From these relations one has ((3.3.64))

$$(3.89)\qquad \begin{aligned} \phi_1(y,\mu,\lambda)\exp\{-iy\,q_+(\mu,\lambda)\} - 1 = \eta_1 + \eta_2 - 1 &= -\int_y^\infty E_{1j}(y',\mu,\lambda)\,\eta_j(y')dy' \\ &- \int_y^\infty \exp\{-2i(y-y')q_+(\mu,\lambda)\} E_{2j}(y',\mu,\lambda)\eta_j(y')dy' \end{aligned}$$

and hence

$$(3.90)\quad |\phi_1(y,\mu,\lambda)\exp\{-iy\,q_+(\mu,\lambda)\} - 1| \le \sum_{j,k=1}^{2}\int_y^\infty |E_{jk}(y',\mu,\lambda)|\,|\eta_k(y')|\,dy'.$$

Using (3.3.66) and the continuity of $B(\mu,\lambda)$ on $\lambda > c^2(-\infty)\mu^2$ it can be shown that

$$(3.91)\qquad \|K(\mu,\lambda)\| \le 1/2 \text{ for } 0 \le \mu \le p_0,\ \omega_0^2 \le \lambda \le \omega_1^2$$

provided $y_1 = y_1(n)$ is large enough. Thus

$$(3.92)\qquad \|\eta\| \le \frac{\eta^0}{1-\|K\|} = \frac{1}{1-\|K\|} \le 2$$

for $0 \le \mu \le p_0$, $\omega_0^2 \le \lambda \le \omega_1^2$. Combining this with (3.90) and (3.3.66) gives

$$(3.93)\qquad \begin{aligned} |\phi_1(y,\mu,\lambda)\exp\{-iy\,q_+(\mu,\lambda)\} - 1| &\le 2\int_y^\infty \sum_{j,k=1}^{2} |E_{jk}(y',\mu,\lambda)|\,dy' \\ &\le C_1\int_y^\infty |\rho(y') - \rho(\infty)|\,dy' \\ &+ C_2\int_y^\infty |c(y') - c(\infty)|\,dy' \end{aligned}$$

for the same values of μ and λ where C_1 and C_2 depends only on p_0, ω_0 and ω_1. It follows from (3.93) and (1.1) that

(3.94) $$\sup |\phi_1(y,|p|,\omega^2) \exp\{-i y q_+(|p|,\omega^2)\} - 1| \le C_3 y^{1-\alpha}$$

for all $y \ge y_1(n)$ where C_3 depends only on p_0, ω_0 and ω_1 (i.e., n). Combining (3.94) with (3.81) gives (3.75).

Proof of Theorem 3.1 (completed). Lemma 3.5 implies that

(3.95) $$\int_{y_0}^{\infty}\int_{B(p_0)} |w(y,p,t)|^2 \, dpdy \le (\pi p_0^2 C^2/2\alpha - 3) y_0^{3-2\alpha}$$

where $3 - 2\alpha < 0$. Thus given any $\varepsilon > 0$ there is a $y_0 = y_0(\varepsilon,n)$ such that

(3.96) $$\int_{y_0}^{\infty}\int_{B(p_0)} |w(y,p,t)|^2 \, dpdy \le \varepsilon$$

for all $t \in R$. Equation (3.77), together with Lemma 3.4 and the estimate (3.96), imply that

(3.97) $$\lim_{t\to\infty} \sup \|w(\cdot,t)\|^2_{L_2(R^3_+(d))} \le \varepsilon .$$

Since $\varepsilon > 0$ is arbitrary this implies (3.71), as required.

The arguments given above, applied to v_ℓ, v_m and v_n, show that the conclusions of Theorem 3.1 hold for all h such that $\hat{h}_- \in C_0(C_+ \cup C_0 \cup C_-)$. Moreover, this set is dense in $L_2(R^3)$ and hence $\Phi_-^* C_0(C_+ \cup C_0 \cup C_-)$ is dense in $\mathcal{H}_f = P_f\mathcal{H}$ by Corollary 9.12 of Chapter 3. These facts can be used to extend (3.62) to all $h \in \mathcal{H}$ because the mappings $U(t) : \mathcal{H}_f \to L_2(R^3)$ and $U_0(t) : \mathcal{H}_f \to L_2(R^3)$ defined by $U(t)h_f = \exp(-i t A^{1/2}) h_f$ and $U_0(t)h_f = v_f^0(t,\cdot)$ are uniformly bounded for all $t \in R$ (see [29, p. 32]). The density argument needed to extend (3.62) to all $h \in \mathcal{H}$ has been given in many places; see, for example, [25, Ch. 2] or [27, p. 260].

§4. TRANSIENT GUIDED WAVES

The asymptotic behavior for $t \to \infty$ of the guided component $u_g(t,\cdot) = P_g u(t,\cdot)$ is derived in this section. $u_g(t,\cdot)$ is a sum in $\mathcal{H}$ of mutually orthogonal partial waves $u_k(t,\cdot) = P_k u(t,\cdot) = \mathrm{Re}\{v_k(t,\cdot)\}$, $1 \le k < N_0$. The starting point for the analysis is the integral representation

(4.1) $$v_k(t,x,y) = \int_{\Omega_k} \psi_k(x,y,p) \exp\{-i t \omega_k(|p|)\} \tilde{h}_k(p) dp$$

where

(4.2) $$\tilde{h}_k(p) = \int_{R^3} \overline{\psi_k(x,y,p)} \, h(x,y) \, c^{-2}(y) \, \rho^{-1}(y) dxdy$$

and the integrals converge in $\mathcal{H}$ and $L_2(\Omega_k)$, respectively. The integral in (4.1) can be written

$$v_k(t,x,y) = \frac{1}{2\pi}\int_{\Omega_k} \exp\{i(x\cdot p - t\omega_k(|p|))\}\,\psi_k(y,|p|)\,\tilde{h}_k(p)\,dp. \tag{4.3}$$

This is an oscillatory integral that can be estimated by the method of stationary phase when $\tilde{h}_k \in C_0^\infty(\Omega_k)$. To apply the method define

$$\begin{cases} r = \sqrt{t^2 + |x|^2} \\ t = r\xi_0,\ x_1 = r\xi_1,\ x_2 = r\xi_2 \\ \xi = (\xi_0,\xi_1,\xi_2) \in S^2 \subset R^3 \end{cases} \tag{4.4}$$

where S^2 denotes the unit sphere in R^3. Then (4.3) takes the form

$$v_k(t,x,y) = \int_{\Omega_k} \exp\{i\,r\,\theta_k(p,\xi)\}\,g_k(p,y)\,dp \tag{4.5}$$

where

$$\begin{cases} \theta_k(p,\xi) = \xi_1 p_1 + \xi_2 p_2 - \xi_0\omega_k(|p|) \\ g_k(p,y) = (2\pi)^{-1}\,\psi_k(y,p)\,\tilde{h}_k(p). \end{cases} \tag{4.6}$$

Estimates for large r of the integral in (4.5) are needed that are uniform for (ξ,y) in compact subsets of $S^2 \times R$. Such estimates are provided by a version of the method of stationary phase due to M. Matsumura [15]. A form of Matsumura's results applicable to (4.5) is presented in Appendix 2. This result is applied below to estimating v_k.

The phase function θ_k has a point of stationary phase if and only if

$$\frac{U_k(|p|)p}{|p|} = \frac{x}{t} \tag{4.7}$$

where

$$U_k(|p|) = \omega_k'(|p|) \tag{4.8}$$

is the group speed associated with the dispersion relation $\omega = \omega_k(|p|)$. It will be assumed, for brevity, that $U_k(\mu)$ is a monotone decreasing function that maps $\mathcal{O}_k$ onto $(c_m, c(\infty))$. In this case (4.7) has a unique solution if $|x|/t$ lies in the range of $U_k(|p|)$; that is

$$c_m < \frac{|x|}{t} < c(\infty), \tag{4.9}$$

and no solution otherwise. The solution is given by

$$p = Q_k(|x|/t)\; x/|x| \tag{4.10}$$

where Q_k is the inverse function to U_k. By calculating the Hessian θ_k'' one can show that

$$r^2 \; |\det \theta_k''(p,\xi)| = t^2 \; U_k(|p|) \; |U_k'(|p|)|/|p| \tag{4.11}$$

and $\operatorname{sgn} \theta_k''(p,\xi) = 0$. In particular, each point of stationary phase is non-degenerate and makes a contribution

$$v_k^\infty(t,x,y,p) = \frac{|p|^{1/2} \exp\{i(|x|\,|p|-t\omega_k(|p|))\}\; \psi_k(y,|p|)\tilde{h}_k(p)}{t\{U_k(|p|)\,|U_k'(|p|)|\}^{1/2}} \tag{4.12}$$

to the interval in (4.3), where p is given by (4.10). For $|x|/t$ outside the interval (4.9) there is no point of stationary phase. Thus the stationary phase approximation to $v_k(t,x,y)$ is given by

$$v_k^\infty(t,x,y) = \chi(|x|/t)\; v_k^\infty(t,x,y,Q_k(|x|/t)\; x/|x|) \tag{4.13}$$

where χ is the characteristic function of the interval $(c_m, c(\infty))$ and one has

Theorem 4.1. For all $h \in \mathcal{H}$ such that $\tilde{h}_k \in C_0^\infty(\Omega_k)$ there exists a constant $C = C_k(h)$ such that

$$|v_k(t,x,y) - v_k^\infty(t,x,y)| \le C/t^2 \tag{4.14}$$

for all $t > 0$, $x \in R^2 - \{0\}$ and $y \in R$.

Theorem 4.1 can be proved by application of Theorems A.1 and A.2 of Appendix 2. The proof that $\psi_k(y,p)$ has the required p-derivatives is lengthy but straightforward and will not be given here. If $\tilde{h}_k$ is not a smooth function then Theorems A.1 and A.2 are not applicable and the estimate (4.14) may fail. However, the definitions (4.12) and (4.13) are meaningful for all $h \in \mathcal{H}$ and one has

Theorem 4.2. For all $h \in \mathcal{H}$, all $t > 0$ and $k = 1,2,3,\cdots$ one has

$$v_k^\infty(t,\cdot) \in \mathcal{H} \tag{4.15}$$

and

$$\|v_k^\infty(t,\cdot)\|_{\mathcal{H}} = \|\tilde{h}_k\|_{L_2(\Omega_k)} = \|P_k h\|_{\mathcal{H}} . \tag{4.16}$$

Moreover, the mapping $t \to v_k(t,\cdot)$ is continuous from R_+ to $\mathcal{H}$ and

$$\lim_{t\to\infty} \|v_k(t,\cdot) - v_k^\infty(t,\cdot)\|_{\mathcal{H}} = 0. \tag{4.17}$$

The proofs of these properties are the same as those for the Pekeris profile, given in [27], and are not reproduced here. On defining

$$u_k^\infty(t,x,y) = \mathrm{Re}\,\{v_k^\infty(t,x,y)\} \tag{4.18}$$

one also has

Corollary 4.3. For all $h \in \mathcal{H}$ and $k = 1,2,3,\cdots$,

$$\lim_{t\to\infty} \|u_k(t,\cdot) - u_k^\infty(t,\cdot)\|_{\mathcal{H}} = 0. \tag{4.19}$$

If $h \in L_2^1(R^3)$ then $u_k(t,\cdot) \in L_2^1(R^3)$ and asymptotic wave functions for the first derivatives of u_k can be constructed. Indeed, if $\tilde{h}_k \in C_0^\infty(\Omega_k)$ then the first derivatives of v_k are given by

$$D_t v_k(t,x,y) = \frac{1}{2\pi}\int_{\Omega_k} \exp\,\{i(x\cdot p - t\omega_k(|p|))\}(-i\omega_k(|p|))\psi_k(y,p)\tilde{h}_k(p)\,dp,$$

(4.20)

$$D_j v_k(t,x,y) = \frac{1}{2\pi}\int_{\Omega_k} \exp\,\{i(x\cdot p - t\omega_k(|p|))(ip_j)\psi_k(y,p)\tilde{h}_k(p)\,dp \quad (j = 1,2),$$

$$D_y v_k(t,x,y) = \frac{1}{2\pi}\int_{\Omega_k} \exp\,\{i(x\cdot p - t\omega_k(|p|))\}\, D_y\psi_k(y,p)\tilde{h}_k(p)\,dp.$$

These integrals have the same form as the integral (4.3) for v_k. The corresponding asymptotic wave functions are defined by

$$v_{k0}^{\infty}(t,x,y,p) = (-i\omega_k(|p|))\, v_k^{\infty}(t,x,y,p),$$

(4.21)

$$v_{kj}^{\infty}(t,x,y,p) = (ip_j)\, v_k^{\infty}(t,x,y,p) \quad (j = 1,2),$$

$$v_{k3}^{\infty}(t,x,y,p) = D_y v_k^{\infty}(t,x,y,p), \text{ and}$$

$$v_{kj}^{\infty}(t,x,y) = \chi(|x|/t)\, v_{kj}^{\infty}(t,x,y,Q_k(|x|/t)\, x/|x|)$$

for $j = 0,1,2,3$. The analogue of Theorem 4.2 is

Theorem 4.4. For all $h \in L_2^1(R^3)$, all $t > 0$ and $k = 1,2,3,\cdots$ one has

$$v_{kj}^{\infty}(t,\cdot) \in L_2(R^3),\ j = 0,1,2,3, \tag{4.22}$$

$$\|v_{k0}^{\infty}(t,\cdot)\|_{\mathcal{H}}^2 + \sum_{j=1}^{3} \|v_{kj}^{\infty}(t,\cdot)\|_{L_2(R^3,\rho^{-1}dxdy)}^2 = 2\, \|A^{1/2}\, h_k\|_{\mathcal{H}}^2, \tag{4.23}$$

and

$$\lim_{t\to\infty} \|D_j v_k(t,\cdot) - v_{kj}^{\infty}(t,\cdot)\|_{L_2(R^3)} = 0,\ j = 0,1,2,3. \tag{4.24}$$

The proof of Theorem 4.4 is the same as for the special case of the Pekeris profile which was treated in detail in [27].

The preceding discussion was restricted to the special case where $U_k(\mu)$ is monotonic. If $U_k(\mu)$ has a finite number of maxima and minima there are a corresponding number of points of stationary phase and the form of the asymptotic wave function is more complicated but still tractable. In the case of the Pekeris profile, treated in [27], there are two points of stationary phase. Cases that lead to infinitely many stationary points have not yet been encountered. They would require additional analysis.

§5. ASYMPTOTIC ENERGY DISTRIBUTIONS

The total energy

$$E(u,R^3,t) = \int_{R^3} \{|\nabla u(t,X)|^2 + c^{-2}(y)|D_0 u(t,X)|^2\}\rho^{-1}(y)dX \tag{5.1}$$

of an arbitrary solution wFE is finite and constant. Moreover, A is the selfadjoint operator in $\mathcal{H}$ associated with the sesquilinear form $\mathcal{A}$ on $\mathcal{H}$

defined by $D(A) = L_2^1(R^3) \subset \mathcal{H}$ and

$$(5.2) \qquad A(u,v) = \int_{R^3} \nabla\bar{u}(X) \cdot \nabla v(X)\, \rho^{-1}(y)\,dX.$$

It follows from Kato's second representation theorem that $D(A^{1/2}) = L_2^1(R^3)$ and for all $u \in D(A^{1/2})$ one has

$$(5.3) \qquad \|A^{1/2}\, u\|_{\mathcal{H}}^2 = A(u,u) = \int_{R^3} |\nabla u(X)|^2 \rho^{-1}(y)\,dX.$$

Hence the total energy satisfies

$$(5.4) \qquad E(u,R^3,t) = \|A^{1/2}\, u\|_{\mathcal{H}}^2 + \|D_0 u\|_{\mathcal{H}}^2.$$

Moreover, if $h \in D(A^{1/2})$ and

$$(5.5) \qquad u(t,X) = \mathrm{Re}\,\{v(t,X)\},\ v(t,\cdot) = e^{-itA^{1/2}}\, h$$

then a simple calculation gives

$$(5.6) \qquad E(u,R^3,t) = \|A^{1/2}\, h\|_{\mathcal{H}}^2.$$

Indeed, $A^{1/2}\, h = A^{1/2}\, f + ig$ where, by assumption, f and g are real-valued. Now $A^{1/2}$ is a real operator; i.e., $A^{1/2}\,\bar{h} = \overline{A^{1/2}\, h}$. It follows that $A^{1/2}\, f$ and g are the real and imaginary parts of $A^{1/2}\, h$, respectively, and (5.6) follows immediately since

$$(5.7) \qquad |A^{1/2}\, h(X)|^2 = |A^{1/2}\, f(X)|^2 + |g(X)|^2.$$

The total energy (5.4) is constant for all $t \geq 0$. The same is true for the partial waves u_f, u_g and u_k, $1 \leq k < N_0$. Moreover, as shown in §2, $\{P_f, P_1, P_2, \cdots\}$ is a complete family of orthogonal projections in $\mathcal{H}$ that reduces A. The energy partition theorem follows immediately:

$$(5.8) \qquad \|A^{1/2}\, h\|_{\mathcal{H}}^2 = \|A^{1/2}\, h_f\|_{\mathcal{H}}^2 + \sum_{k=1}^{N_0-1} \|A^{1/2}\, h_k\|_{\mathcal{H}}^2.$$

The partial energies may be calculated from the initial state by

$$(5.9) \qquad E(u_f,R^3,t) = \|A^{1/2}\, h_f\|_{\mathcal{H}}^2 = \int_{R^3} \omega^2(P)\ |\hat{h}_-(P)|^2 dP$$

and

$$E(u_k, R^3, t) = \|A^{1/2} h_k\|^2_{\mathcal{H}} = \int_{\Omega_k} \omega_k^2(|p|)|\tilde{h}_k(p)|^2 dp. \tag{5.10}$$

The results on asymptotic energy distributions, formulated above as (1.45)-(1.55), can now be proved. The results on the free component u_f follow immediately from Corollary 3.3 and the results on free waves proved in detail in [25]. The results on the guided components u_k follow from Theorem 4.4. These results were proved in detail in [27] for the case of the Pekeris operator. The proofs are identical for the class of operators treated here and therefore will not be repeated.

§6. SEMI-INFINITE AND FINITE LAYERS

The preceding analysis is extended in this section to the cases of semi-infinite and finite layers of stratified fluid. The extensions are based on the normal mode expansions for these cases that were derived in Chapter 3, §10. Only the principal concepts and results are formulated here since the proofs are entirely analogous to those of the preceding sections.

Semi-Infinite Layers. As in Chapter 3, §10, the fluid is assumed to occupy the domain R^3_+ and to satisfy the Dirichlet or Neumann boundary condition. Here the functions $\rho(y)$ and $c(y)$ are assumed to be Lebesgue measurable and satisfy

$$0 < \rho_m \leq \rho(y) \leq \rho_M < \infty, \quad 0 < c_m \leq c(y) \leq c_M < \infty \tag{6.1}$$

and

$$|\rho(y) - \rho(\infty)| \leq C\, y^{-\alpha}, \quad |c(y) - c(\infty)| \leq C\, y^{-\alpha} \tag{6.2}$$

for all $y > 0$ where ρ_m, ρ_M, $\rho(\infty)$, c_m, c_M, $c(\infty)$, C and α are constants and

$$\alpha > 3/2. \tag{6.3}$$

As in Chapter 3, §10, the acoustic propagators for $\rho(y)$ and $c(y)$ corresponding to the Dirichlet and Neumann conditions will be denoted by A^0 and A^1, respectively. They are selfadjoint non-negative linear operators in $\mathcal{H}_+ = L_2(R^3_+, c^{-2}(y)\rho^{-1}(y)dxdy)$.

The normal mode functions $\psi^j(x,y,p,\lambda)$ for A^j, as defined in Chapter 3, §10 are parameterized by $(p,\lambda) \in \Omega = \{(p,\lambda) \mid \lambda > c^2(\infty)|p|^2\}$. Their asymptotic form for $y \to \infty$ is given by

(6.4) $$\psi^j(x,y,p,q) \sim \frac{c(|p|,\lambda)}{2\pi}\left\{e^{i(p\cdot x-yq)} + R^j e^{i(p\cdot x+yq)}\right\}$$

where $q = q(|p|,\lambda) = (\lambda\, c^{-2}(\infty) - |p|^2)^{1/2}$, $c(|p|,\lambda) = (\rho(\infty)/4\pi q(|p|,\lambda))^{1/2}$, $R^j = R^j(|p|,\lambda)$ and $|R^j(|p|,\lambda)| = 1$. As in the preceding sections, it will be convenient to introduce new parameters: $(p,\lambda) \to (p,q) = (p,q(|p|,\lambda)) \in R^3_+$ and normal mode functions

(6.5) $$\phi^j_+(x,y,p,q) = (2q)^{1/2}\, c(\infty)\, \psi^j(x,y,p,\lambda)$$

where

(6.6) $$\lambda = \lambda(p,q) = c^2(\infty)(|p|^2 + q^2).$$

The asymptotic form of ϕ^j_+ is

(6.7) $$\phi^j_+(x,y,p,q) \sim \frac{c(\infty)\rho^{1/2}(\infty)}{(2\pi)^{3/2}}\left\{e^{i(p\cdot x-qy)} + R^j e^{i(p\cdot x+qy)}\right\},\quad y \to \infty.$$

The second family ϕ^j_- defined by

(6.8) $$\phi^j_-(x,y,p,q) = \overline{\phi^j_+(x,y,-p,q)}$$

is also needed. It satisfies

(6.9) $$\phi^j_-(x,y,p,q) \sim \frac{c(\infty)\rho^{1/2}(\infty)}{(2\pi)^{3/2}}\left\{e^{i(p\cdot x+qy)} + \overline{R^j}\, e^{i(p\cdot x-qy)}\right\},\quad y \to \infty.$$

The expansion theorem of Chapter 3, §10 implies that the limits

(6.10) $$\hat{f}^j_\pm(p,q) = L_2(R^3_+)\text{-}\lim_{M\to\infty} \int_0^M \int_{|x|\le M} \overline{\phi^j_\pm(x,y,p,q)}\, f(x,y)\, c^{-2}(y)\rho^{-1}(y)\,dx\,dy$$

exist. Moreover, if

(6.11) $$\Phi^j_\pm : \mathcal{H} \to L_2(R^3_+)$$

is defined by $\Phi^j_\pm f = \hat{f}^j_\pm$ then $\Phi^j_\pm$ is a partial isometry with range $L_2(R^3_+)$ and

$$\Phi_\pm^{j*}\,\Phi_\pm^j + \sum_{k=1}^{N_0^j-1} \Psi_k^{j*}\,\Psi_k^j = 1 \tag{6.12}$$

where $\Psi_k^j : \mathcal{H} \to L_2(\Omega_k)$ are the partial isometries associated with the guided wave normal modes $\psi_k^j(x,y,p)$ of Chapter 3, §10.

Normal mode expansions for A^j are given by (6.12) with either the + or - sign. (6.12) implies that the orthogonal projections in $\mathcal{H}_+$ defined by

$$\begin{cases} P_f^j = \Phi_+^{j*}\,\Phi_+^j = \Phi_-^{j*}\,\Phi_-^j \\ P_k^j = \Psi_k^{j*}\,\Psi_k^j, \; 1 \le k < N_0^j \end{cases} \tag{6.13}$$

form a complete family that reduces A^j.

Transient Free Waves for Semi-Infinite Layers. The free component of a complex acoustic potential

$$v(t,\cdot) = \exp\,(-i\,t\,(A^j)^{1/2})\,h,\; h \in \mathcal{H}_+, \tag{6.14}$$

is given by

$$v_f(t,\cdot) = P_f^j v(t,\cdot) = \exp\,(-i\,t\,(A^j)^{1/2})\,P_f^j h. \tag{6.15}$$

The ϕ_-^j-representation of v_f is

$$v_f(t,x,y) = \int_{R_+^3} \phi_-^j(x,y,p,q)\,\exp\,(-i\,t\,\omega(p,q))\,\hat{h}_-^j(p,q)\,dp\,dq \tag{6.16}$$

where $\omega(p,q) = c(\infty)\sqrt{|p|^2 + q^2}$. Moreover, as in §3 one can write

$$\phi_-^j(x,y,p,q) = \frac{c(\infty)\rho^{1/2}(\infty)}{(2\pi)^{3/2}}\left\{e^{i(p\cdot x+qy)}\,I_-^j(y,p,q) + e^{i(p\cdot x-qy)}R_-^j(y,p,q)\right\} \tag{6.17}$$

where

$$\begin{cases} \lim_{y\to\infty} I_-^j(y,p,q) = 1, \\ \lim_{y\to\infty} R_-^j(y,p,q) = \overline{R^j(p,\lambda)}. \end{cases} \tag{6.18}$$

Then, proceeding as in §3, one can prove the following analogue of Theorem 3.1.

Theorem 6.1. For every $h \in \mathcal{H}_+$ let $v_f^0(t,\cdot)$ be defined by

$$v_f^0(t,x,y) = \exp\,(-i\,t\,c(\infty)A_0^{1/2})\,h_0(x,y), \quad (x,y) \in R_+^3 \tag{6.19}$$

where $h_0 \in L_2(R^3)$ is the function whose Fourier transform is

$$\hat{h}_0(p,q) = \begin{cases} c(\infty)\rho^{1/2}(\infty)\,\hat{h}_-^j(p,q), & (p,q) \in R_+^3, \\ 0, & (p,q) \in R_-^3. \end{cases} \tag{6.20}$$

Then

$$\lim_{t\to\infty} \|v_f(t,\cdot) - v_f^0(t,\cdot)\|_{\mathcal{H}_+} = 0. \tag{6.21}$$

Transient Guided Waves. For both semi-infinite and finite layers the form of the guided components $v_k(t,\cdot)$ is precisely the same as for the case of an infinite layer. Thus the analysis of §4 applies unchanged to these cases.

Chapter 5
Scattering of Signals by Inhomogeneous Layers

The purpose of this chapter is to study the scattering and distortion of acoustic signals by an inhomogeneous plane-stratified layer separating two homogeneous fluids. The transmitter or source of the signals is assumed to lie in one of the homogeneous fluids and the reflected and transmitted signals are calculated. The physical hypotheses and results are summarized in §1. The proofs, which are based on the preceding chapters are developed in the remaining sections.

§1. SUMMARY

Throughout this chapter it is assumed that $\rho(y)$ and $c(y)$ satisfy the boundedness conditions (2.1.4) and are variable only in a finite layer $y_1 \le y \le y_2$. More precisely, it is assumed that

$$\begin{cases} \rho(y) = \rho(-\infty) \text{ and } c(y) = c(-\infty) \text{ for } y < y_1, \\ \\ \rho(y) = \rho(\infty) \text{ and } c(y) = c(\infty) \text{ for } y > y_2. \end{cases} \tag{1.1}$$

These conditions imply, in a trivial way, that $\rho(y)$ and $c(y)$ satisfy the hypotheses of the preceding chapters.

The scattering by the inhomogeneous layer $y_1 \le y \le y_2$ of signals whose sources are localized in the half-space $y > y_2$ will be analyzed. As described in Chapter 2, this can be modelled by initial values $u(0,\cdot) = f$, $D_0 u(0,\cdot) = g$ which satisfy

$$\text{supp } f \cup \text{supp } g \cup \{X : x_1^2 + x_2^2 + (y - y_0)^2 \le \delta^2\}. \tag{1.2}$$

The sources will lie in the region $y > y_2$ if $y_0 > y_2 + \delta$.

The Source Radiation Pattern. The incident signal will be defined to be the acoustic field that would be generated by the given sources if they were situated in an unlimited homogeneous fluid with parameters $\rho(\infty)$, $c(\infty)$. It is characterized by the potential $u_{inc}(t,X)$ that satisfies (2.1.6) and (2.1.8) with $\rho(y) = \rho(\infty)$ and $c(y) = c(\infty)$ everywhere. Of particular interest for applications is the far field form of the incident signal. It is described by the asymptotic wave functions [25,28]

$$u^{\infty}_{inc,k}(t,X) = r^{-1}\, s_k(r - c(\infty)t,\Theta), \quad k = 0,1,2,3, \tag{1.3}$$

where

$$r = |X| = \sqrt{x_1^2 + x_2^2 + y^2} \tag{1.4}$$

and

$$\Theta = \frac{X}{r} \in S^2. \tag{1.5}$$

S^2 is the unit sphere (= set of all unit vectors) in R^3. It was shown in Chapter 4 that for suitable functions s_k,

$$D_k u_{inc}(t,\cdot) = u^{\infty}_{inc,k}(t,\cdot) + o_k, \quad k = 0,1,2,3, \tag{1.6}$$

where $o_k \to 0$ in $L_2(R^3)$ when $t \to \infty$. In addition, one has [25]

$$s_k(\tau,\Theta) = -c^{-1}(\infty)\, \Theta_k s_0(\tau,\Theta), \; k = 1,2,3. \tag{1.7}$$

Hence, the far field form of the incident signal is characterized by the single real-valued function $s_0(\tau,\Theta)$, defined for $\tau \in R$, $\Theta \in S^2$. Moreover, for incident fields wFE one has [25] $s_0 \in L_2(R \times S^2)$ and

$$2c^{-2}(\infty)\rho^{-1}(\infty)\, \|s_0\|^2_{L_2(R\times S^2)} = \int_{R^3} \{|\nabla f|^2 + c^{-2}(\infty)|g|^2\}\rho^{-1}(\infty)\,dxdy \tag{1.8}$$

is the total signal energy. This is verified in §2 below.

The function $s_0(\tau,\Theta)$ will be called the source radiation pattern. It is uniquely determined by the transmitter or sources through the initial state (f,g). The exact relationship is given below. However, it is the values of $s_0(\tau,\Theta)$, rather than $f(X)$ and $g(X)$, that are the primary data of the signal scattering problem studied here. Indeed s_0 is directly observable through the relations (1.3)-(1.7). Moreover, it can be shown that a signal wFE in a homogeneous fluid that is generated in any way by sources confined to a bounded region will have the asymptotic behavior (1.3)-(1.7). In applications the design of a pulse mode transmitter with a prescribed radiation pattern s_0 is a primary goal of the design engineer.

The Source Momentum Distribution. The Fourier transform of s_0,

$$\hat{s}_0(\omega,\Theta) = \frac{1}{(2\pi)^{1/2}} \int_{-\infty}^{\infty} e^{-i\omega\tau} s_0(\tau,\Theta)d\tau, \tag{1.9}$$

converges in $L_2(R \times S^2)$ (cf. [25]) and satisfies

$$\hat{s}_0(-\omega,\Theta) = \overline{\hat{s}_0(\omega,\Theta)} \text{ for all } \omega \geq 0 \tag{1.10}$$

if and only if $s_0(\tau,\Theta)$ is real-valued. Hence the transforms of the source radiation patterns are characterized by their values for positive frequencies ω. For these values it is shown in §2 below that $\hat{s}_0$ is related to the momentum distribution of the sources by

$$\hat{s}_0(\omega,\Theta) = \frac{-i\omega}{2} \hat{h}_0(\omega\Theta) \tag{1.11}$$

where $h_0 = g - i\, c(\infty)A_0^{1/2} f \in L_2(R^3)$ (for notation see (4.3.17)) and $\hat{h}_0$ is the usual Fourier transform in $L_2(R^3)$:

$$\hat{h}_0(P) = \frac{1}{(2\pi)^{3/2}} \int_{R^3} e^{-iP\cdot X} h_0(X)dX, \tag{1.12}$$

with $P = (p,q) = (p_1,p_2,q) \in R^3$. $\hat{h}_0(P)$ is the complex amplitude associated with the signal momentum P. It is related to the initial state by

$$\hat{h}_0(P) = \hat{g}(P) - i\omega_+(P)\, \hat{f}(P) \tag{1.13}$$

where $\omega_\pm(P) = c(\pm\infty)|P|$, as in Chapter 4.

s_0 is determined by the momentum distribution $\hat{h}_0$ through (1.9)-(1.11). Conversely, if s_0 is known then $\hat{h}_0$ can be recovered by (1.9) and

$$\hat{h}_0(P) = 2i|P|^{-1}\, \hat{s}_0(|P|,|P|^{-1}P). \tag{1.14}$$

Moreover,

$$\int_{R^3} |\hat{h}_0(P)|^2 dP = 4 \int_0^\infty \int_{S^2} |\hat{s}_0(\omega,\Theta)|^2 d\Theta d\omega = 2\, \|s_0\|^2_{L_2(R\times S^2)}. \tag{1.15}$$

Thus the correspondence $\sqrt{2}\, s_0 \to \hat{h}_0$ defines a unitary mapping of the Hilbert space of real radiation patterns wFE onto the Hilbert space of all momentum distributions in $L_2(R^3)$. The quantity (1.15) is proportional to the total energy of the incident signal (cf. (1.8)).

The function $\hat{h}_0$ will be called the source momentum distribution. In what follows it will be convenient to describe the reflected and transmitted signals by their momentum distributions, rather than their radiation patterns. The latter can always be recovered from the relations (1.9)-(1.11).

The Structure of the Scattered Signal. The total acoustic field wFE $u(t,X)$ generated by the sources in the presence of the plane-stratified scattering layer has a decomposition

$$u = u_{\text{free}} + u_{\text{guided}} \tag{1.16}$$

where $u_{\text{free}} = P_f u$ and $u_{\text{guided}} = P_g u$. The structure of u_{guided} was analyzed thoroughly in Chapter 4 and is not discussed further here. The component u_{free} satisfies, by (4.1.50),

$$E^{\infty}(u_{\text{free}}, R^2 \times [y_1, y_2]) = 0. \tag{1.17}$$

Hence the free component of the signal is asymptotically negligible in the scattering layer.

The reflected and transmitted signals will be defined for $X \in R^2_+(y_2)$ and $X \in R^2_-(y_1)$, respectively, by

$$u_{\text{refl}}(t,X) = u_{\text{free}}(t,X) - u_{\text{inc}}(t,X),\ y \geq y_2, \tag{1.18}$$

$$u_{\text{trans}}(t,X) = u_{\text{free}}(t,X),\ y \leq y_1. \tag{1.19}$$

These functions satisfy wave equations for homogeneous fluids with sound speeds $c(\infty)$ and $c(-\infty)$, respectively, and have finite energy. Moreover, the results of Chapter 4 imply that they have asymptotic wave functions:

$$D_0 u_{\text{refl}}(t,X) = r^{-1}\, s_{\text{refl}}(r - c(\infty)t, \Theta) + o_{\text{refl}} \tag{1.20}$$

$$D_0 u_{\text{trans}}(t,X) = r^{-1}\, s_{\text{trans}}(r - c(-\infty)t, \Theta) + o_{\text{trans}} \tag{1.21}$$

where $o_{\text{refl}} \to 0$ and $o_{\text{trans}} \to 0$ in $L_2(R^3_+(y_2))$ and $L_2(R^3_-(y_1))$, respectively, when $t \to \infty$. Note that s_{refl} and s_{trans} lie in the complementary subspaces of $L_2(R \times S^2)$ defined by $L_2(R \times S_+^2)$ and $L_2(R \times S_-^2)$, respectively, where $S_\pm^2 = S^2 \cap R_\pm^2$.

The goal of this chapter is to calculate the radiation patterns s_{refl} and s_{trans} and to determine how they vary with the signal radiation pattern s_0 and the fluid parameters $\rho(y)$ and $c(y)$. This will be done by calculating their momentum distributions

(1.22) $$\hat{h}_{refl}(P) = 2i|P|^{-1} s_{refl}(|P|,|P|^{-1}P), \quad P \in R_+^3,$$

(1.23) $$\hat{h}_{trans}(P) = 2i|P|^{-1} s_{trans}(|P|,|P|^{-1}P), \quad P \in R_-^3.$$

To describe the results it will be convenient to introduce the mappings in momentum space defined by

(1.24) $$\Pi_R : R_+^3 \to R_-^3, \ \Pi_R(p,q) = (p,-q),$$

and

(1.25) $$\Pi_T : R_-^3 \to -C_+, \ \Pi_T(p,q) = (p,-q_+(|p|,\lambda(p,q))).$$

(See Chapter 3 for the definitions of C_+ and q_+.) Π_R obviously defines the reflection of wave momentum in the plane $q = 0$. It is easy to verify that

(1.26) $$c(\infty)|\Pi_T(P)| = c(-\infty)|P| \text{ for all } P \in R_-^3.$$

(1.25) and (1.26) imply that the momenta $\Pi_T(P)$ and P are related by Snell's law for the refraction of a plane wave passing from a medium with propagation speed $c(\infty)$ to one with a speed $c(-\infty)$.

With the above notation the principal result of this chapter takes the form

(1.27) $$\hat{h}_{refl}(P) = \begin{cases} R_+(|p|,\lambda(P))\, \hat{h}_0(\Pi_R(P)), & P \in C_+, \\ R_0(|p|,\lambda(P))\, \hat{h}_0(\Pi_R(P)), & P \in C_0, \end{cases}$$

and

$$(1.28) \qquad \hat{h}_{\text{trans}}(P) = \frac{c^2(-\infty)\rho(-\infty)}{c^2(\infty)\rho(\infty)} T_-(|p|,\lambda(P))\, \hat{h}_0(\Pi_T(P)),\ P \in C_-,$$

where R_+, R_0 and T_- are the reflection and transmission coefficients of the normal mode function $\phi_+(x,y,p,q)$ associated with $\rho(y)$ and $c(y)$ as in Chapter 3.

Equation (1.27) states that the complex amplitude $\hat{h}_{\text{refl}}(P)$ associated with momentum $P \in C_+$ (resp., $P \in C_0$) is the product of R_+ (resp., R_0) and the amplitude $\hat{h}_0(\Pi_R(P))$ associated with the momentum $\Pi_R(P) \in R^3$ Similarly, (1.28) states that the amplitude $\hat{h}_{\text{trans}}(P)$ associated with momentum $P \in C_-$ is the product of $c^2(-\infty)\rho(-\infty)c^{-2}(\infty)\rho^{-1}(\infty)T_-$ and the amplitude $\hat{h}_0(\Pi_T(P))$ associated with the momentum $\Pi_T(P) \in -C_+$.

Relationship to the Scattering Operator. There is a close relationship between the scattering relations (1.27), (1.28) and the scattering operator S of the stratified fluid layer characterized by $\rho(y)$ and $c(y)$. S is the unitary operator in $L_2(R^3)$ defined by

$$(1.29) \qquad S = \Phi_- \Phi_+^*$$

where $\Phi_\pm : \mathcal{H}_f \to L_2(R^3)$ are the normal mode mappings of Chapter 3, §9. It is shown below that S is determined by the coefficients R_+, R_0, R_-, T_+, T_- associated with $\phi_+(X,P)$ and the mapping

$$(1.30) \qquad \Pi : C_+ \to C_- = R_-^3$$

defined by

$$(1.31) \qquad \Pi(p,q) = (p,-q_-(|p|,\lambda(p,q)).$$

Π^{-1} exists and is given by

$$(1.32) \qquad \Pi^{-1}(p,q) = (p,q_+(|p|,\lambda(p,q))).$$

With this notation the construction of S is given by

$$(1.33) \qquad S\,h(P) = \begin{cases} R_+h(P) + \dfrac{c(\infty)\rho^{1/2}(\infty)}{c(-\infty)\rho^{1/2}(-\infty)} T_+h(\Pi(P)), & P \in C_+, \\ R_0h(P), & P \in C_0, \\ R_-h(P) + \dfrac{c(-\infty)\rho^{1/2}(-\infty)}{c(\infty)\rho^{1/2}(\infty)} T_-h(\Pi^{-1}(P)), & P \in C_-, \end{cases}$$

where $R_\pm = R_\pm(|p|,\lambda(P))$, etc. Relation (1.33) is derived in §4 below. The unitarity of S in $L_2(R^3)$ imposes certain restrictions on $R_\pm$, R_0 and $T_\pm$ which are also derived in §4. In §5 the construction (1.33) is used to derive an alternative representation of s_{refl} and s_{trans}.

§2. SIGNALS IN HOMOGENEOUS FLUIDS

The theory of asymptotic wave functions for sound waves in homogeneous fluids was developed in [25, Ch. 2]. In that work the constant density and sound speed were normalized to have the value unity. Here the theory is needed for arbitrary constant density $\rho(\infty)$ and sound speed $c(\infty)$, for comparison with inhomogeneous stratified fluids. The theory may be obtained as a very simple case of the results of Chapter 4 by the specialization

$$(2.1)\qquad \begin{cases} \rho(y) \to \rho(\infty) = \text{const.}, \\ c(y) \to c(\infty) = \text{const.} \end{cases}$$

The notation $\mathcal{H}(\infty) = L_2(R^3, c^{-2}(\infty)\rho^{-1}(\infty)dxdy)$ and $A(\infty) = -c^2(\infty)\nabla^2$ will be used for the corresponding Hilbert space and acoustic propagator.

The specialization (2.1) implies that $\mathcal{H} = \mathcal{H}(\rho,c) \to \mathcal{H}(\infty) = \mathcal{H}(\infty)_f$; i.e., there are no guided modes. Moreover

$$(2.2)\qquad q_+(\mu,\lambda) = q_-(\mu,\lambda) \to q(\mu,\lambda) = (\lambda\, c^{-2}(\infty) - \mu^2)^{1/2}$$

and

$$(2.3)\qquad \lambda(p,q) \to c^2(\infty)(|p|^2 + q^2) \equiv \omega_+^2(p,q) \text{ for all } (p,q) \in R^3.$$

The normal mode functions $\phi_\pm(X,P)$ of Chapter 3 become plane waves

$$(2.4)\qquad \begin{cases} \phi_+ \to \phi_+^\infty(x,y,p,q) = c(\infty)\rho^{1/2}(\infty)\,(2\pi)^{-3/2}\, e^{i(p\cdot x - qy)}, \\ \phi_- \to \phi_-^\infty(x,y,p,q) = c(\infty)\rho^{1/2}(\infty)\,(2\pi)^{-3/2}\, e^{i(p\cdot x + qy)}, \end{cases}$$

while the normal mode expansions are essentially the Fourier integral representation in $L_2(R^3)$. Thus for all $h \in \mathcal{H}(\infty)$ (isomorphic with $L_2(R^3)$) one has the expansions

$$\hat{h}_{\pm}^{\infty}(p,q) = \frac{c(\infty)\rho^{1/2}(\infty)}{(2\pi)^{3/2}} \int_{R^3} e^{-i(p\cdot x\mp qy)}\, h(x,y)\, c^{-2}(\infty)\rho^{-1}(\infty)\,dxdy \tag{2.5}$$

in $L_2(R^3)$ and

$$h(x,y) = \frac{c(\infty)\rho^{1/2}(\infty)}{(2\pi)^{3/2}} \int_{R^3} e^{i(p\cdot x\mp qy)}\, \hat{h}_{\pm}^{\infty}(p,q)\,dpdq \tag{2.6}$$

in $\mathcal{H}(\infty)$. Clearly

$$\hat{h}_{\pm}^{\infty}(p,q) = c^{-1}(\infty)\rho^{-1/2}(\infty)\, \hat{h}(p,\mp q) \tag{2.7}$$

where $\hat{h}$ is the usual Fourier transform, defined by (1.12), and (2.6) is equivalent to the usual Fourier integral representation. Parseval's relation may be written

$$\|h\|_{\mathcal{H}(\infty)} = \|\hat{h}_{\pm}^{\infty}\|_{L_2(R^3)}. \tag{2.8}$$

Asymptotic Wave Functions. To obtain the asymptotic wave functions for the incident signal u_{inc} note that it is the solution in $\mathcal{H}(\infty)$ of

$$D_0^2 u_{inc} + A(\infty)u_{inc} = 0, \quad t \in R, \tag{2.9}$$

$$u_{inc}(0) = f \text{ and } D_0 u_{inc}(0) = g. \tag{2.10}$$

The solution can be written

$$u_{inc}(t,\cdot) = \cos\,(t\, A^{1/2}(\infty))f + (A^{-1/2}(\infty)\,\sin\, t\, A^{1/2}(\infty))g. \tag{2.11}$$

If, as will be assumed,

$$f \in D(A^{1/2}(\infty)), \quad g \in D(A^{-1/2}(\infty)) \tag{2.12}$$

then u_{inc} is a solution wFE and

$$u_{inc}(t,X) = \text{Re}\,\{v_{inc}(t,X)\} \tag{2.13}$$

where

$$v_{inc}(t,X) = \int_{R^3} \phi_\pm^\infty(X,P)\, e^{-it\omega_+(P)}\, \hat{h}_\pm^\infty(P)\,dP \tag{2.14}$$

$$= \frac{c(\infty)\rho^{1/2}(\infty)}{(2\pi)^{3/2}} \int_{R^3} e^{i(p\cdot x \mp qy - t\omega_+(P))}\, \hat{h}_\pm^\infty(p,q)\,dp\,dq$$

and

$$\hat{h}_\pm^\infty(P) = \hat{f}_\pm^\infty(P) + i\omega_+^{-1}(P)\, \hat{g}_\pm^\infty(P). \tag{2.15}$$

On choosing the lower signs in (2.14) and using (2.7) this becomes the Fourier representation

$$v_{inc}(t,X) = \frac{1}{(2\pi)^{3/2}} \int_{R^3} e^{i(X\cdot P - c(\infty)t|P|)}\, \hat{h}(P)\,dP \tag{2.16}$$

with

$$\hat{h}(P) = \hat{f}(P) + i\omega_+^{-1}(P)\, \hat{g}(P). \tag{2.17}$$

Note that this is identical with [25, (2.34)] with t replaced by $c(\infty)t$. Thus the asymptotic wave functions for v_{inc} may be obtained from those of [25, Ch. 2] by the same substitution. In particular, one has

$$v_{inc}(t,\cdot) \sim v_{inc}^\infty(t,\cdot) \text{ in } \mathcal{H}(\infty) \text{ when } t \to \infty \tag{2.18}$$

where

$$v_{inc}^\infty(t,X) = r^{-1}\, G(r - c(\infty)t, \Theta), \tag{2.19}$$

$$G(\tau,\Theta) = \frac{1}{(2\pi)^{1/2}} \int_0^\infty e^{i\tau\omega}\, \hat{h}(\omega\Theta)(-i\omega)\,d\omega \tag{2.20}$$

in $L_2(R \times S^2)$ and

$$\|G\|_{L_2(R\times S^2)} = \|h\|_{L_2(R^3)}. \tag{2.21}$$

Derivatives. Hypotheses (2.12) imply that v_{inc} has first derivatives in $L_2(R^3)$ that may be obtained from (2.16) by differentiation under the integral sign. These integrals also have the form (2.16) and proceeding as in [25, Ch. 2] gives

$$D_k v_{inc}(t,X) \sim v_{inc,k}^{\infty}(t,X) = r^{-1} G_k(r - c(\infty)t,\Theta) \tag{2.22}$$

in $L_2(R^3)$ when $t \to \infty$ where

$$G_0(\tau,\Theta) = \frac{1}{(2\pi)^{1/2}} \int_0^{\infty} e^{i\tau\omega} \hat{h}_0(\omega\Theta)(-i\omega)\,d\omega, \tag{2.23}$$

$$\hat{h}_0(\omega\Theta) \equiv -i\, c(\infty)\omega\, \hat{h}(\omega\Theta) = \hat{g}(\omega\Theta) - i\, c(\infty)\omega\, \hat{f}(\omega\Theta), \tag{2.24}$$

and

$$G_k(\tau,\Theta) = -c^{-1}(\infty)\Theta_k\, G_0(\tau,\Theta),\ k = 1,2,3. \tag{2.25}$$

On taking real parts one obtains the results (1.3)-(1.7) and (1.9)-(1.13) quoted in §1.

Asymptotic Energy Distributions. The total energy of the incident signal is

$$\begin{aligned} E(u_{inc},R^3,0) &= E(u_{inc},R^3,t) \\ &= \int_{R^3} \{|\nabla u_{inc}(t,X)|^2 + c^{-2}(\infty)|D_0 u_{inc}(t,X)|^2\}\rho^{-1}(\infty)\,dX. \end{aligned} \tag{2.26}$$

It may be expressed in terms of s_0 by combining (1.3)-(1.7) and (2.26) (cf. [25, Ch. 8]). The result is

$$E(u_{inc},R^3,0) = 2\, c^{-2}(\infty)\rho^{-1}(\infty)\, \|s_0\|^2_{L_2(R\times S^2)} \tag{2.27}$$

which verifies (1.8). Asymptotic energy distributions in cones and other subsets may be derived by the same method, as in [25, Ch. 8].

§3. THE REFLECTED AND TRANSMITTED SIGNALS

The functions s_{refl} and s_{trans} are calculated in this section. First, general representations of s_{refl} and s_{trans} are derived that are valid under the hypotheses of Chapter 4. Next it is shown that for finite scattering layers the asymptotic forms for $y \to \infty$ of the normal mode functions is exact outside the scattering layer. These results are then combined to verify the representations (1.27), (1.28) for finite scattering layers. The implications of these results for signal distortion are discussed at the end of the section.

For simplicity it is assumed in the remainder of the report that $f \in D(A^{1/2})$ and $g \in D(A^{-1/2})$ so that the representation $u(t,X) = \text{Re}\,\{v(t,X)\}$, $v(t,\cdot) = \exp(-itA^{1/2})h$ is available; cf. [25, Ch. 3]. All of the results obtained may be extended, by a density argument, to arbitrary fields wFE; i.e., $f \in D(A^{1/2}) = L_2^1(R^3)$ and $g \in L_2(R^3)$, but the details will not be given here.

The starting point of the calculations is Chapter 4, Corollary 3.3. The hypotheses on f and g imply that

$$(3.1) \qquad u_{\text{free}}(t,X) = \text{Re}\,\{v_{\text{free}}(t,X)\}$$

and for $k = 0,1,2,3$

$$(3.2) \qquad D_k v_{\text{free}}(t,X) \sim D_k v_{\text{free}}^0(t,X) \text{ in } L_2(R^3) \text{ when } t \to \infty$$

where

$$(3.3) \qquad v_{\text{free}}^0(t,\cdot) = \begin{cases} e^{-itA^{1/2}(\infty)}\, h^+ \text{ in } L_2(R_+^3(y_2)), \\ e^{-itA^{1/2}(-\infty)}\, h^- \text{ in } L_2(R_-^3(y_1)), \end{cases}$$

and h^+ and h^- are the functions in $L_2(R^3)$ whose Fourier transforms are (Chapter 4, Theorem 3.1)

$$(3.4) \qquad \hat{h}^+(P) = \begin{cases} c(\infty)\rho^{1/2}(\infty)\,\hat{h}_-(P), & P \in R_+^3, \\ 0, & P \in R_-^3, \end{cases}$$

$$(3.5) \qquad \hat{h}^-(P) = \begin{cases} 0, & P \in R_+^3, \\ c(-\infty)\rho^{1/2}(-\infty)\,\hat{h}_-(P), & P \in R_-^3. \end{cases}$$

Recall that $h = f + iA^{-1/2}g$ and hence

$$(3.6) \qquad \hat{h}_-(P) = \hat{f}_-(P) + i\lambda^{-1/2}(P)\,\hat{g}_-(P).$$

The Reflected Signal. Applying the same formalism to the incident signal gives

$$u_{inc}(t,X) = \mathrm{Re}\,\{v_{inc}(t,X)\} \tag{3.7}$$

where

$$v_{inc}(t,\cdot) = e^{-itA^{1/2}(\infty)}\, h_{inc} \tag{3.8}$$

and $h_{inc} = f + iA^{-1/2}(\infty)\, g$, whence

$$\hat{h}_{inc}(P) = \hat{f}(P) + i\omega_+^{-1}(P)\, \hat{g}(P). \tag{3.9}$$

Subtracting (3.7) from (3.1) gives (Definition (1.18))

$$u_{refl}(t,X) = \mathrm{Re}\,\{v_{refl}(t,X)\} \tag{3.10}$$

where

$$v_{refl}(t,\cdot) = v_{free}(t,\cdot) - v_{inc}(t,\cdot). \tag{3.11}$$

It follows from (3.2), (3.3) that

$$D_k v_{refl}(t,X) \sim D_k v^0_{refl}(t,X) \text{ in } L_2(R^3_+(y_2)),\ t \to \infty, \tag{3.12}$$

where

$$v^0_{refl}(t,\cdot) = v^0_{free}(t,\cdot) - v_{inc}(t,\cdot) = e^{-itA^{1/2}(\infty)}\,(h^+ - h_{inc}). \tag{3.13}$$

In particular,

$$D_0 v_{refl}(t,X) \sim e^{-itA^{1/2}(\infty)}\, h_{refl} \text{ in } L_2(R^3_+(y_2)),\ t \to \infty, \tag{3.14}$$

where

$$h_{refl} = -iA^{1/2}(\infty)\,(h^+ - h_{inc}). \tag{3.15}$$

Equations (3.14), (3.15) imply the validity of (1.20) with (cf. (2.16)-(2.20))

(3.16) $$\hat{s}_{\text{refl}}(\omega,\Theta) = \frac{(-i\omega)}{2}\,\hat{h}_{\text{refl}}(\omega\Theta).$$

On taking the Fourier transform of (3.15), using (3.4), (3.6) and (3.9) and recalling that $\lambda(P) = \omega_+^2(P)$ when $P \in R_+^3$, one finds that the momentum distribution of s_{refl} is

$$\hat{h}_{\text{refl}}(P) = (-i\omega_+(P))[c(\infty)\rho^{1/2}(\infty)\,\hat{h}_-(P) - \hat{h}_{\text{inc}}(P)]$$

(3.17)

$$= c(\infty)\rho^{1/2}(\infty)\,\hat{g}_-(P) - \hat{g}(P) - i\omega_+(P)[c(\infty)\rho^{1/2}(\infty)\,\hat{f}_-(P) - \hat{f}(P)],\ P \in R_+^3.$$

Using (2.7) this can be written

(3.18) $$\hat{h}_{\text{refl}}(P) = c(\infty)\,\rho^{1/2}(\infty)\,[\hat{g}_-^{sc}(P) - i\omega_+(P)\hat{f}_-^{sc}(P)],\ P \in R_+^3,$$

where

(3.19) $$\hat{g}_-^{sc}(P) = \hat{g}_-(P) - \hat{g}_-^{\infty}(P)$$

and similarly for $\hat{f}_-^{sc}$.

<u>The Transmitted Signal</u>. Proceeding similarly gives

(3.20) $$u_{\text{trans}}(t,X) = u_{\text{free}}(t,X) = \text{Re}\,\{v_{\text{free}}(t,X)\}$$

where, by (3.2)-(3.6), one has

(3.21) $$D_0 v_{\text{free}}(t,X) \sim e^{-itA^{1/2}(-\infty)}\,h_{\text{trans}} \text{ in } L_2(R_-^3(y_1)),\ t \to \infty,$$

with

(3.22) $$h_{\text{trans}} = -iA^{1/2}(-\infty)\,h^-.$$

These relations imply the validity of (1.21) with

(3.23) $$s_{\text{trans}}(\omega,\Theta) = \frac{(-i\omega)}{2}\,\hat{h}_{\text{trans}}(\omega\Theta)$$

defined by

$$(3.24) \qquad \hat{h}_{trans}(P) = (-i\omega_-(P))\ c(-\infty)\ \rho^{1/2}(-\infty)\ \hat{h}_-(P),\ P \in R^3_-.$$

Using (3.6) with $\lambda(P) = \omega_-^2(P)$ for $P \in R^3_-$ gives the representation

$$(3.25) \qquad \hat{h}_{trans}(P) = c(-\infty)\ \rho^{1/2}(-\infty)\ [\hat{g}_-(P) - i\omega_-(P)\ \hat{f}_-(P)],\ P \in R^3_-.$$

The representations (3.18) and (3.25) are valid for the class of stratified fluids defined in Chapter 4. They are used below to derive (1.27) and (1.28).

A Decomposition of the Normal Mode Functions. It will be convenient to write the normal mode functions $\phi_\pm$ as

$$(3.26) \qquad \begin{cases} \phi_+(X,P) = \phi_{in}(X,P) + \phi_+^{sc}(X,P), \\ \\ \phi_-(X,P) = \phi_{out}(X,P) + \phi_-^{sc}(X,P), \end{cases}$$

where

$$(3.27) \qquad \phi_{in}(X,P) = \begin{Bmatrix} (2\pi)^{-3/2}\ c(\infty)\rho^{1/2}(\infty)\ e^{i(p\cdot x-qy)}, & P \in R^3_+ \\ \\ 0, & P \in R^3_- \end{Bmatrix},\ X \in R^3_+,$$

$$(3.28) \qquad \phi_{in}(X,P) = \begin{Bmatrix} 0, & P \in R^3_+ \\ \\ (2\pi)^{-3/2}\ c(-\infty)\rho^{1/2}(-\infty)\ e^{i(p\cdot x-qy)}, & P \in R^3_- \end{Bmatrix},\ X \in R^3_-,$$

and

$$(3.29) \qquad \phi_{out}(x,y,p,q) = \overline{\phi_{in}(x,y,-p,q)}.$$

Note that ϕ_{out} is given by equations (3.27), (3.28) with q replaced by -q. The behavior of $\phi_\pm(X,P)$ for $y \to \pm\infty$, which was determined in Chapter 3, may be described by the equations

$$(3.30)\qquad \phi_+^{sc}(X,P) = c(P)\begin{Bmatrix} R_+\ e^{iP\cdot X}, & P\in C_+ \\ R_0\ e^{iP\cdot X}, & P \in C_0 \\ T_-\ e^{i(p\cdot x+q_+y)}, & P\in C_- \end{Bmatrix} + \sigma_+(X,P)$$

$$(3.31)\qquad \phi_+^{sc}(X,P) = c(P)\begin{Bmatrix} T_+\ d^{i(p\cdot x-q_-y)}, & P\in C_+ \\ T_0\ e^{ip\cdot x}\ e^{q'_-y}, & P \in C_0 \\ R_-\ e^{ip\cdot X}, & P\in C_- \end{Bmatrix} + \sigma_-(X,P)$$

where $R_+ = R_+(|p|,\lambda(P))$, etc., $q_\pm = q_\pm(|p|,\lambda(P))$, $q'_- = q'_-(|p|,\lambda(P))$, $c(P)$ is defined by (3.1.44), and

$$(3.32)\qquad \sigma_\pm(x,y,p,q) \to 0 \text{ when } y \to \pm\infty.$$

Similar statements for ϕ_-^{sc} follow from the relation $\phi_-^{sc}(x,y,p,q) = \overline{\phi_+^{sc}(x,y,-p,q)}$. The derivation of (1.27) and (1.28) for the case of finite scattering layers is based on

Lemma 3.1. If $\rho(y)$ and $c(y)$ are constant outside the interval $[y_1,y_2]$ then $\sigma_+(x,y,p,q) = 0$ for $y > y_2$ and $\sigma_-(x,y,p,q) = 0$ for $y < y_1$.

The vanishing of $\sigma_+(x,y,p,q)$ for $y > y_2$ when $\rho(y) = \rho(\infty)$, $c(y) = c(\infty)$ for $y > y_2$ follows from the observation that (3.30) with $\sigma_+(x,y,p,q) = 0$ is a solution of the equation $A\phi_+^{sc}(X,P) = \lambda(P)\ \phi_+^{sc}(X,P)$ for $y > y_2$ together with the unique continuation property of the ordinary differential equation for the normal mode function $\psi_+(x,y,p,\lambda) = (2\pi)^{-1}\ e^{ip\cdot x}\ \psi_+(y,|p|,\lambda)$, etc. (see (3.3.1)). The same argument applies to σ_- on $y < y_1$.

Verification of (1.27) for Finite Layers. The case where $P \in C_+$ will be treated, the other case being similar. Assume that $\rho(y)$ and $c(y)$ are constant outside $[y_1,y_2]$ and that $\operatorname{supp} f \subset R_+^3(y_2)$, so that $\rho(y) = \rho(\infty)$ and $c(y) = c(\infty)$ on supp f. Then

$$(3.33)\qquad \begin{aligned} \hat{f}_-(P) &= \int_{R^3} \overline{\phi_-(X,P)}\ f(X)\ c^{-2}(y)\rho^{-1}(y)\,dX \\ &= c^{-2}(\infty)\rho^{-1}(\infty)\left\{\int_{R^3} \overline{\phi_{out}(X,P)}\ f(X)\,dX + \int_{R^3} \overline{\phi_-^{sc}(X,P)}\ f(X)\,dX\right\}. \end{aligned}$$

Substituting $\overline{\phi_-^{sc}(x,y,p,q)} = \phi_+^{sc}(x,y,-p,q)$ from (3.30) with $P \in C_+$ and

$\sigma_+ = 0$ (Lemma 3.1) gives

$$\begin{aligned}\hat{f}_-(P) &= c^{-2}(\infty)\rho^{-1}(\infty)\Big[c(\infty)\rho^{1/2}(\infty)\,(2\pi)^{-3/2}\int_{R^3} e^{-iP\cdot X} f(X)\,dX \\ &\quad + c(\infty)\rho^{1/2}(\infty)\,(2\pi)^{-3/2}\int_{R^3} R_+\, e^{-i(p\cdot x-qy)} f(X)\,dX\Big] \\ &= c^{-1}(\infty)\rho^{-1/2}(\infty)\,(\hat{f}(P) + R_+\,\hat{f}(p,-q)).\end{aligned} \tag{3.34}$$

Moreover, $\hat{f}_-^{\infty}(P) = c^{-1}(\infty)\rho^{-1/2}(\infty)\,\hat{f}(P)$, by (2.7). Subtracting this equation from (3.34) gives

$$\begin{aligned}\hat{f}_-^{sc}(P) &= c^{-1}(\infty)\rho^{-1/2}(\infty)\,R_+\,\hat{f}(p,-q) \\ &= c^{-1}(\infty)\rho^{-1/2}(\infty)\,R_+(|p|,\lambda(P))\,\hat{f}(\Pi_R(P))\end{aligned} \tag{3.35}$$

for every $P \in C_+$. The same relation holds for $\hat{g}_-^{sc}$. Substituting these relations into (3.18) and using the definition (2.24) of $\hat{h}_0$ gives (1.27).

Verification of (1.28) for Finite Layers. If $P \in C_-$, $\rho(y)$ and $c(y)$ are constant outside $[y_1,y_2]$ and supp $f \subset R_+^3(y_2)$ one has by the analogous argument, using Lemma 3.1,

$$\begin{aligned}\hat{f}_-(P) &= c^{-2}(\infty)\rho^{-1}(\infty)\int_{R^3} \overline{\phi_-(X,P)}\, f(X)\,dX \\ &= c^{-2}(\infty)\rho^{-1}(\infty)c(-\infty)\rho^{1/2}(-\infty)\,(2\pi)^{-3/2}\,T_-\int_{R^3} e^{-i(p\cdot x-q_+y)} f(X)\,dX \\ &= c^{-2}(\infty)\rho^{-1}(\infty)c(-\infty)\rho^{1/2}(-\infty)\,T_-\,\hat{f}(p,-q_+) \\ &= c^{-2}(\infty)\rho^{-1}(\infty)c(-\infty)\rho^{1/2}(-\infty)\,T_-\,\hat{f}(\Pi_T(P)).\end{aligned} \tag{3.36}$$

The analogous equation holds for $\hat{g}_-(P)$. Substituting these relations into (3.25) and noting that for $P \in C_-$ one has $\omega_+(\Pi_T(p)) = c(\infty)|\Pi_T(P)| = c(-\infty)|P| = \omega_-(P)$, by (1.34), gives (1.28). This completes the proof of

Theorem 3.2. If $\rho(y)$ and $c(y)$ are constant outside an interval $[y_1,y_2]$ and if supp $f \cup$ supp $g \subset R_+^3(y_2)$ then the reflected and transmitted signal momentum distributions are given by (1.27) and (1.28).

Signal Distortion. Combining (1.22) and (1.27) gives

$$
\begin{aligned}
\hat{s}_{\text{refl}}(\omega,\Theta) &= -\frac{i\omega}{2}\hat{h}_{\text{refl}}(\omega\Theta) \\
&= -\frac{i\omega}{2} R_{+}(\omega \sin\theta, c^2(\infty)\omega^2)\, \hat{h}_0(\Pi_R(\omega\Theta))
\end{aligned} \tag{3.37}
$$

for $\Theta \in C_+$ where

$$\Theta = (\Theta_1,\Theta_2,\Theta_3) = (\cos\phi \sin\theta, \sin\phi \sin\theta, \cos\theta). \tag{3.38}$$

Defining $\Theta_R \in R^3_-$ by

$$\Pi_R(\omega\Theta) = \omega(\Theta_1,\Theta_2,-\Theta_3) = \omega\,\Theta_R, \tag{3.39}$$

(3.37) can be written

$$\hat{s}_{\text{refl}}(\omega,\Theta) = R_{+}(\omega \sin\theta, c^2(\infty)\omega^2)\hat{s}_0(\omega,\Theta_R), \quad \Theta \in C_+, \tag{3.40}$$

and similarly

$$\hat{s}_{\text{refl}}(\omega,\Theta) = R_0(\omega \sin\theta, c^2(\infty)\omega^2)\, \hat{s}_0(\omega,\Theta_R), \quad \Theta \in C_0. \tag{3.41}$$

In the same way, beginning with (1.23) and (1.28) one finds

$$
\begin{aligned}
\hat{s}_{\text{trans}}(\omega,\Theta) &= -\frac{i\omega}{2}\hat{h}_{\text{trans}}(\omega\Theta) \\
&= -\frac{i\omega}{2}\frac{c^2(-\infty)\rho(-\infty)}{c^2(\infty)\rho(\infty)} T_{-}(\omega \sin\theta, c^2(-\infty)\omega^2)\, \hat{h}_0(\Pi_T(\omega\Theta))
\end{aligned} \tag{3.42}
$$

for $\Theta \in C_-$. Now, by (1.26),

$$|\Pi_T(\omega\Theta)| = \gamma^{-1}\omega, \quad \gamma = \frac{c(\infty)}{c(-\infty)}\,. \tag{3.43}$$

Hence (3.42) can be written

$$\hat{s}_{\text{trans}}(\omega,\Theta) = \frac{c^2(-\infty)\rho(-\infty)}{c^2(\infty)\rho(\infty)} T_{-}(\omega \sin\theta, c^2(-\infty)\omega^2)\, \hat{s}_0(\gamma^{-1}\omega,\Theta_T) \tag{3.44}$$

where $\Theta_T \in C_-$ is defined by

$$\Pi_T(\omega\Theta) = \gamma^{-1}\,\omega\,\Theta_T. \tag{3.45}$$

Equations (3.40), (3.41) and the Fourier inversion formula imply the representations

$$s_{\text{refl}}(\omega,\Theta) = \text{Re}\left\{\left(\frac{2}{\pi}\right)^{1/2}\int_0^\infty e^{i\tau\omega}\, R_+(\omega\sin\theta, c^2(\infty)\omega^2)\, \hat{s}_0(\omega,\Theta_R)\,d\omega\right\},\ \Theta\in C_+,$$

(3.46)

$$= \text{Re}\left\{\left(\frac{2}{\pi}\right)^{1/2}\int_0^\infty e^{i\tau\omega}\, R_0(\omega\sin\theta, c^2(\infty)\omega^2)\, \hat{s}_0(\omega,\Theta_R)\,d\omega\right\},\ \Theta\in C_0.$$

Similarly, (3.44) and the Fourier inversion formula, followed by a simple change of variable, give

$$s_{\text{trans}}(\omega,\Theta) = \frac{c(-\infty)\rho(-\infty)}{c(\infty)\rho(\infty)}\,\text{Re}\left\{\left(\frac{2}{\pi}\right)^{1/2}\int_0^\infty e^{i\gamma\tau\omega}\, T_-(\gamma\omega\sin\theta, c^2(\infty)\omega^2)\hat{s}_0(\omega,\Theta_T)\,d\omega\right\}$$
(3.47)

for $\Theta \in C_-$. These results show explicitly how the signal distortion due to reflection and transmission by a scattering layer is determined by the reflection and transmission coefficients of the normal mode function $\phi_+(X,P)$.

§4. CONSTRUCTION OF THE SCATTERING OPERATOR

The scattering operator $S : L_2(R^3) \to L_2(R^3)$ is defined by

$$S = \Phi_-\,\Phi_+^* \tag{4.1}$$

where $\Phi_\pm : \mathcal{H}_f \to L_2(R^3)$ are the operators associated in Chapter 3 with the two normal mode expansions in $\mathcal{H}_f$ defined by ϕ_+ and ϕ_-. The unitarity of Φ_+ and Φ_-, proved in Chapter 3, implies that S is unitary and, in particular

$$S^{-1} = S^* = \Phi_+\,\Phi_-^*. \tag{4.2}$$

The defining relations

$$\hat{h}_\pm = \Phi_\pm h \text{ for all } h \in \mathcal{H}_f \tag{4.3}$$

imply that the functions $\hat{h}_\pm \in L_2(R^3)$ are related by

$$\hat{h}_- = S\,\hat{h}_+,\ \hat{h}_+ = S^*\,\hat{h}_-. \tag{4.4}$$

These relations will be used to calculate S.

It was shown in Chapter 4 that, for all $h \in \mathcal{H}_f$,

$$e^{-itA^{1/2}} h \sim e^{-itA^{1/2}(\infty)} h^+ \text{ in } L_2(R^3_+),\ t \to \infty, \tag{4.5}$$

$$e^{-itA^{1/2}} h \sim e^{-itA^{1/2}(-\infty)} h^- \text{ in } L_2(R^3_-),\ t \to \infty, \tag{4.6}$$

where h^+ and h^- are the functions in $L_2(R^3)$ whose Fourier transforms are defined by

$$\hat{h}^+(P) = \begin{cases} c(\infty)\rho^{1/2}(\infty)\,\hat{h}_-(P), & P \in R^3_+, \\ 0, & P \in R^3_-, \end{cases} \tag{4.7}$$

and

$$\hat{h}^-(P) = \begin{cases} 0, & P \in R^3_+, \\ c(-\infty)\rho^{1\ 2}(-\infty)\,\hat{h}_-(P), & P \in R^3_-, \end{cases} \tag{4.8}$$

respectively. Moreover, it is clear from the results of [25] that $\hat{h}^+$ and $\hat{h}^-$ are uniquely determined in $L_2(R^3_+)$ and $L_2(R^3_-)$, respectively, by (4.5) and (4.6). The relationship between $\hat{h}_-$ and $\hat{h}_+$ will now be determined by calculating $\hat{h}^+$ and $\hat{h}^-$ in a second way, using the ϕ_+-representation for $\exp(-itA^{1/2})\,h$, and equating the resulting representations to those given by (4.7), (4.8).

<u>Second Calculation of $\hat{h}^\pm$ Based on ϕ_+</u>. The starting point is the representation (4.3.2):

$$\begin{aligned}(e^{-itA^{1/2}} h)(X) &= \int_{R^3} \phi_+(X,P)\ e^{-it\lambda^{1/2}(P)}\ \hat{h}_+(P)\,dP \\ &= v_a(t,X) + v_b(t,X) + v_c(t,X)\end{aligned} \tag{4.9}$$

where (cf. Chapter 4, §3 for notation)

$$\begin{cases} v_a = e^{-itA^{1/2}}\ \Phi_+^*\ a \\ v_b = e^{-itA^{1/2}}\ \Phi_+^*\ b \\ v_c = e^{-itA^{1/2}}\ \Phi_+^*\ c \end{cases} \tag{4.10}$$

and

$$a = \chi_+ \hat{h}_+, \; b = \chi_0 \hat{h}_+, \; c = \chi_- \hat{h}_+. \tag{4.11}$$

The asymptotic behavior of the components v_a, v_b and v_c will be calculated by the method of Chapter 4, §3. Only the formal steps of the calculations will be given. They can be justified by the method of Chapter 4.

Behavior of v_a. v_a has the representation

$$v_a(t,X) = \int_{C_+} \phi_+(X,P) \exp(-it\omega_+(P))\, a(P)\,dP \tag{4.12}$$

where (3.1.39)

$$\phi_+(x,y,p,q) = (2\pi)^{-1} c(\infty)(2q)^{1/2} e^{ip\cdot x} \psi_+(y,|p|,\lambda) \tag{4.13}$$

for $(p,q) = X_+(p,\lambda) \in C_+$ (and hence $\lambda = \lambda(p,q) = c^2(\infty)(|p|^2 + q^2)$). On employing the representations (4.3.9) and (4.3.11) for ψ_+, noting that $q_+(|p|,\lambda(p,q)) = q$ for $P \in C_+$, and proceeding as in Chapter 4, §3 one finds

$$v_a(t,\cdot) \sim \begin{cases} v_a^1(t,\cdot) \text{ in } L_2(R_+^3), \\ v_a^2(t,\cdot) \text{ in } L_2(R_-^3), \end{cases} \tag{4.14}$$

for $t \to \infty$ where

$$\begin{cases} v_a^1(t,\cdot) = e^{-itA^{1/2}(\infty)} h_a^1 \\ v_a^2(t,\cdot) = e^{-itA^{1/2}(-\infty)} h_a^2 \end{cases} \tag{4.15}$$

are defined by

$$v_a^1(t,X) = c(\infty)\rho^{1/2}(\infty)\,(2\pi)^{-3/2} \int_{C_+} e^{i(X\cdot P - t\omega_+(P))} R_+(|p|,\lambda(P))\, \hat{h}_+(P)\,dP \tag{4.16}$$

and

$$v_a^2(t,X) = c(\infty)\rho^{1/2}(\infty)\,(2\pi)^{-3/2} \int_{C_+} e^{i(x\cdot p - yq_-(|p|,\lambda) - t\omega_+(P))} \times T_+(|p|,\lambda(P))\, \hat{h}_+(P)\,dP. \tag{4.17}$$

To see that v_a has the form of (4.15) note that the mapping $(p,q) \to (p,q')$ $= \Pi(p,q)$ defined by (1.31) maps C_+ onto C_-, has Jacobian $\partial(p,q)/\partial(p,q')$ $= c^2(-\infty)q'/c^2(\infty)q$ and satisfies $\lambda(\Pi(P)) = \lambda(P)$. Changing the variables of integration in (4.17) from $P = (p,q)$ to (p,q') and using (1.32) for Π^{-1} gives

$$v_a^2(t,X) = \frac{c(\infty)\rho^{1/2}(\infty)}{(2\pi)^{3/2}} \int_{C_-} e^{i(x\cdot p+yq'-t\omega_-(p,q'))} T_+(|p|,\lambda(p,q')) \times \hat{h}_+(\Pi^{-1}(p,q')) \frac{c^2(-\infty)}{c^2(\infty)} \left(\frac{-q'}{q_+}\right) dpdq' \tag{4.18}$$

where $q_+ = q_+(|p|,\lambda(p,q'))$.

Behavior of v_b. v_b has the representation

$$v_b(t,X) = \int_{C_0} \phi_+(X,P) \exp(-it\omega_+(P)) b(P)dP \tag{4.19}$$

where

$$\phi_+(x,y,p,q) = (2\pi)^{-1} c(\infty)(2q)^{1/2} e^{ip\cdot x} \psi_0(y,|p|,\lambda) \tag{4.20}$$

for $(p,q) \in X_0(p,\lambda) \in C_0$ (and hence $\lambda = \lambda(p,q) = \omega_+^2(p,q)$). On employing the representations (4.3.29) and (4.3.31) for ψ_0 and proceeding as before one finds

$$v_b(t,\cdot) \sim \begin{cases} v_b^1(t,\cdot) & \text{in } L_2(R_+^3), \\ 0 & \text{in } L_2(R_-^3), \end{cases} \tag{4.21}$$

for $t \to \infty$ where

$$v_b^1(t,\cdot) = e^{-itA^{1/2}(\infty)} h_b^1 \tag{4.22}$$

is defined by

$$v_b^1(t,X) = \frac{c(\infty)\rho^{1/2}(\infty)}{(2\pi)^{3/2}} \int_{C_0} e^{i(X\cdot P-t\omega_+(P))} R_0(|p|,\lambda(P)) \hat{h}_+(P)dP. \tag{4.23}$$

Behavior of v_c. v_c has the representation

$$v_c(t,X) = \int_{C_-} \phi_+(X,P) \exp(-it\omega_-(P)) c(P)dP \tag{4.24}$$

where

$$\phi_+(x,y,p,q) = (2\pi)^{-1}\, c(-\infty)(-2q)^{1/2}\, e^{ip\cdot x}\, \psi_-(y,|p|,\lambda) \tag{4.25}$$

for $(p,q) = X_-(p,\lambda) \in C_-$ (and hence $\lambda = \lambda(p,q) = \omega_-^2(p,q)$). On employing the representations (4.3.43) and (4.3.45) for ψ_- and proceeding as before one finds

$$v_c(t,\cdot) \sim \begin{cases} v_c^1(t,\cdot) \text{ in } L_2(R_+^3), \\ v_c^2(t,\cdot) \text{ in } L_2(R_-^3), \end{cases} \tag{4.26}$$

for $t \to \infty$ where

$$\begin{cases} v_c^1(t,\cdot) = e^{-itA^{1/2}(\infty)}\, h_c^1 \\ v_c^2(t,\cdot) = e^{-itA^{1/2}(-\infty)}\, h_c^2 \end{cases} \tag{4.27}$$

are defined by

$$v_c^1(t,X) = \frac{c(-\infty)\rho^{1/2}(-\infty)}{(2\pi)^{3/2}} \int_{C_-} e^{i(x\cdot p+yq_+-t\omega_-(P))}\, T_-(|p|,\lambda)\, \hat{h}_+(P)dP \tag{4.28}$$

and

$$v_c^2(t,X) = \frac{c(-\infty)\rho^{1/2}(-\infty)}{(2\pi)^{3/2}} \int_{C_-} e^{i(X\cdot P-t\omega_-(P))}\, R_-(|p|,\lambda(P))\, \hat{h}_+(P)dP. \tag{4.29}$$

In (4.28), $q_+ = q_+(|p|,\lambda(p,q))$ and $\lambda = \lambda(p,q)$. To see that v_c^1 has the form of (4.27), recall that the mapping $(p,q) \to (p,q') = (p,q_+(|p|,\lambda(p,q)))$ $= \Pi^{-1}(p,q)$ (see (1.32)). Thus changing to the variables (p,q') in (4.28) gives

$$\begin{aligned} v_c^1(t,X) &= \frac{c(-\infty)\rho^{1/2}(-\infty)}{(2\pi)^{3/2}} \int_{C_+} e^{i(x\cdot p+yq'-t\omega_+(p,q'))} \\ &\quad \times T_-(|p|,\lambda(p,q'))\, \hat{h}_+(\Pi(p,q'))\, \frac{c^2(\infty)}{c^2(-\infty)}\, \frac{q'}{q_-}\, dpdq' \end{aligned} \tag{4.30}$$

where $q_- = q_-(|p|,\lambda(p,q'))$.

Combining the above results gives

$$e^{-itA^{1/2}} h \sim v_a^1(t,\cdot) + v_b^1(t,\cdot) + v_c^1(t,\cdot) \text{ in } L_2(R_+^3), \tag{4.31}$$

$$e^{-itA^{1/2}} h \sim v_a^2(t,\cdot) + 0 + v_c^2(t,\cdot) \text{ in } L_2(R_-^3), \tag{4.32}$$

when $t \to \infty$. Comparing this with (4.5), (4.6) and using the uniqueness of $\hat{h}^+$ in $L_2(R_+^3)$ and of $\hat{h}^-$ in $L_2(R_-^3)$ gives

$$\begin{aligned} \hat{h}^+(P) &= c(\infty)\rho^{1/2}(\infty)\, R_+(|p|,\lambda)\, \hat{h}_+(P) \\ &+ c(-\infty)\rho^{1/2}(-\infty)\, T_-(|p|,\lambda)\, \hat{h}_+(\Pi(P))\, \frac{c^2(\infty)}{c^2(-\infty)}\, \frac{q}{q_-} \text{ in } C_+ \end{aligned} \tag{4.33}$$

$$h^+(P) = c(\infty)\rho^{1/2}(\infty)\, R_0(|p|,\lambda)\, \hat{h}_+(P) \text{ in } C_0 \tag{4.34}$$

while

$$\begin{aligned} \hat{h}^-(P) &= c(\infty)\rho^{1/2}(\infty)\, T_+(|p|,\lambda)\, \hat{h}_+(\Pi^{-1}(P))\, \frac{c^2(-\infty)}{c^2(\infty)}\, \frac{(-q)}{q_+} \\ &+ c(-\infty)\rho^{1/2}(-\infty)\, R_-(|p|,\lambda)\, \hat{h}_+(P) \text{ in } C_-. \end{aligned} \tag{4.35}$$

Comparing these representations with (4.7), (4.8) gives

$$\hat{h}_-(P) = \begin{cases} R_+\, \hat{h}_+(P) + \dfrac{c(\infty)\rho^{1/2}(-\infty)}{c(-\infty)\rho^{1/2}(\infty)}\, T_-\, \hat{h}_-(\Pi(P))\, \dfrac{q}{q_-} & \text{in } C_+, \\ R_0\, \hat{h}_+(P) & \text{in } C_0, \\ R_-\, \hat{h}_+(P) + \dfrac{c(-\infty)\rho^{1/2}(\infty)}{c(\infty)\rho^{1/2}(-\infty)}\, T_+\, \hat{h}_+(\Pi^{-1}(P))\, \dfrac{(-q)}{q_+} & \text{in } C_-, \end{cases} \tag{4.36}$$

where $R_+ = R_+(|p|,\lambda(P))$, etc.

Relation (4.36), together with (4.4) provides a representation of S which is not identical with (1.33). However, (4.36) and the unitarity of S implies certain relations among the coefficients $R_\pm$, R_0 and $T_\pm$. These may be obtained directly from (4.36) and the integral identity $\|\hat{h}_-\| = \|\hat{h}_+\|$. However, they are more easily obtained from the bracket operation of Lagrange's formula, as in Chapter 3, §5. There the relations

(4.37) $$\frac{q_\pm}{\rho(\pm\infty)} |R_\pm|^2 + \frac{q_\mp}{\rho(\mp\infty)} |T_\pm|^2 = \frac{q_\pm}{\rho(\pm\infty)}, \ \lambda \in \Lambda,$$

(4.38) $$|R_0| = 1, \ \lambda \in \Lambda_0$$

were derived in this way. Similarly, calculating the limits of $[\psi_+,\psi_-]$ for $y \to \pm\infty$ gives

(4.39) $$\frac{q_+}{\rho(\infty)} T_- = \frac{q_-}{\rho(-\infty)} T_+, \ \lambda \in \Lambda,$$

where $q_\pm = q_\pm(\mu,\lambda)$, $T_\pm = T_\pm(\mu,\lambda)$. In particular, for $P = (p,q) \in C_+$, $q_+(|p|,\lambda(P)) = q$ and

(4.40) $$T_-(|p|,\lambda(P)) \frac{q}{q_-} = \frac{\rho(\infty)}{\rho(-\infty)} T_+(|p|,\lambda(P))$$

while for $P = (p,q) \in C_-$, $q_-(|p|,\lambda(P)) = -q$ and

(4.41) $$T_+(|p|,\lambda(P)) \frac{(-q)}{q_+} = \frac{\rho(-\infty)}{\rho(\infty)} T_-(|p|,\lambda(P)).$$

Combining (4.36), (4.40) and (4.41) yields the representation (1.33) for S.

§5. THE SCATTERING OPERATOR AND SIGNAL STRUCTURE

To describe the relationship between S and the scattered signals s_{refl} and s_{trans} it will be convenient to introduce the orthogonal projection Q_- in $L_2(R^3)$ defined by

(5.1) $$Q_- h = \chi_- h$$

where χ_- is the characteristic function of $C_- = R^3_-$, as before. Note that one may rewrite (1.27) as

(5.2) $$\hat{h}_{refl}(P) = R_+(|p|,\lambda(P)) \, Q_- \hat{h}_0(\Pi_R(P)), \ P \in C_+$$

because $P \in C_+ \Rightarrow \Pi_R(P) \in C_-$. If the unitary operator $R : L_2(R^3) \to L_2(R^3)$ defined by

(5.3) $$R\, h(P) = h(\Pi_R(P)) \text{ for all } P \in R^3$$

is used then (5.2) is equivalent to

$$\hat{h}_{refl}(P) = R_+(|p|,\lambda(P))(R\,Q_-\hat{h}_0)(P),\ P \in C_+. \tag{5.4}$$

Finally, on applying the representation (1.33) of S to the function $h = RQ_-\hat{h}_0$ and noting that $\text{supp}\, RQ_-\hat{h}_0 \subset R^3_+$ it is seen that

$$\hat{h}_{refl}(P) = SRQ_-\hat{h}_0(P) \text{ for all } P \in C_+. \tag{5.5}$$

The same calculation with $P \in C_0$ gives

$$\hat{h}_{refl}(P) = SRQ_-\hat{h}_0(P) \text{ for all } P \in C_0. \tag{5.6}$$

Now consider (1.28). It can be rewritten

$$\hat{h}_{trans}(P) = \frac{c^2(-\infty)\rho(-\infty)}{c^2(\infty)\rho(\infty)}\, T_-(|p|,\lambda(P))\, Q_-\hat{h}_0(\Pi_T(P)),\ P \in C_-. \tag{5.7}$$

Note that definitions (1.24), (1.25) and (1.31) of Π_R, Π_T and Π imply that

$$\Pi_T = \Pi_R\, \Pi^{-1}. \tag{5.8}$$

Substituting in (5.7) and proceeding as before gives

$$\hat{h}_{trans}(P) = \frac{c^2(-\infty)\rho^{1/2}(-\infty)}{c^2(\infty)\rho^{1/2}(\infty)}\, T_-(|p|,\lambda(P))\, RQ_-\hat{h}_0(\Pi^{-1}(P)),\ P \in C_-. \tag{5.9}$$

Finally, applying the representation (1.33) of S to $h = RQ_-\hat{h}_0$ and $P \in C_-$ gives, since $\text{supp}\, RQ_-\hat{h}_0 \subset R^3_+$,

$$\hat{h}_{trans}(P) = \frac{c(-\infty)\rho^{1/2}(-\infty)}{c(\infty)\rho^{1/2}(\infty)}\, SRQ_-\hat{h}_0(P) \text{ for all } P \in C_-. \tag{5.10}$$

The representations (5.5), (5.6) and (5.10) reveal that the scattered signal depends only on s_0 and S. Indeed, the occurrence of $Q_-\hat{h}_0$ implies that only the values of $s_0(\tau,\Theta)$ with $\Theta \in S^2_-$ are needed to find the scattered signals. Of course, these facts depend on the assumption that the signal sources lie in $R^3_+(y_2)$. If they lie in $R^3_-(y_1)$ then a similar relation holds with $Q_-\hat{h}_0$ replaced by $Q_+\hat{h}_0 = (\chi_+ + \chi_0)h$. If sources occur in the scattering layer then the scattered signals cannot be calculated using only S.

Relations (5.5), (5.6) and (5.10) can be given a more symmetrical form by noting that (3.18) and (3.25) can be written

$$(5.11)\quad \begin{cases} \hat{h}_{\text{refl}}(P) = c(\infty)\rho^{1/2}(\infty)\,\hat{k}_-^{sc}(P), & P \in R_+^3 \\ \hat{h}_{\text{trans}}(P) = c(-\infty)\rho^{1/2}(-\infty)\,\hat{k}_-^{sc}(P), & P \in R_-^3 \end{cases}$$

where

$$(5.12)\quad k = (-iA^{1/2})\,h = g - i\,A^{1/2}\,f.$$

It follows that, for supp f ∪ supp g ⊂ $R_+^3(y_2)$, relations (5.5), (5.6) and (5.10) can be combined as

$$(5.13)\quad \hat{k}_-^{sc}(P) = c^{-1}(\infty)\rho^{-1/2}(\infty)\,SRQ_-\hat{h}_0(P),\ P \in R^3.$$

Energy in the Scattered Field. The total energy in the scattered signal field is given by the limit

$$(5.14)\quad E_{sc}^{\infty} = E^{\infty}(u_{\text{refl}}, R_+^3(y_2)) + E^{\infty}(u_{\text{trans}}, R_-^3(y_1))$$

because of (1.17). E_{sc}^{∞} can be calculated from the incident signal by means of (1.20), (1.21) and (5.11), (5.12). Indeed, proceeding as in [25] one finds that

$$(5.15)\quad \begin{aligned} E^{\infty}(u_{\text{refl}}, R_+^3(y_2)) &= 2c^{-2}(\infty)\rho^{-1}(\infty)\,\|\hat{s}_{\text{refl}}\|^2_{L_2(R\times S_+^2)} \\ &= 2c^{-2}(\infty)\rho^{-1}(\infty)\int_{-\infty}^{\infty}\int_{S_+^2}|\hat{s}_{\text{refl}}(\omega,\Theta)|^2\,d\Theta d\omega \\ &= 4c^{-2}(\infty)\rho^{-1}(\infty)\int_{0}^{\infty}\int_{S_+^2}|\hat{s}_{\text{refl}}(\omega,\Theta)|^2\,d\Theta d\omega \\ &= c^{-2}(\infty)\rho^{-1}(\infty)\int_{0}^{\infty}\int_{S_+^2}|\hat{h}_{\text{refl}}(\omega\Theta)|^2\,\omega^2\,d\Theta d\omega \\ &= c^{-2}(\infty)\rho^{-1}(\infty)\,\|\hat{h}_{\text{refl}}\|^2_{L_2(R_+^3)} \\ &= \|\hat{k}_-^{sc}\|^2_{L_2(R_+^3)} \end{aligned}$$

and, by a similar calculation,

(5.16)
$$\mathcal{E}^{\infty}(u_{trans}, R^3_-(y_1)) = \|\hat{k}^{sc}_-\|^2_{L_2(R^3_-)},$$

whence

(5.17)
$$\mathcal{E}^{\infty}_{sc} = \|\hat{k}^{sc}_-\|^2_{L_2(R^3)}.$$

Combining this with (5.13) and using the unitarity of S and R gives

(5.18)
$$\begin{aligned}\mathcal{E}^{\infty}_{sc} &= c^{-2}(\infty)\rho^{-1}(\infty)\|Q_-\hat{h}_0\|^2_{L_2(R^3)} \\ &= c^{-2}(\infty)\rho^{-1}(\infty)\int_{R^3_-}|\hat{h}_0(P)|^2\, dP.\end{aligned}$$

By using the relation (2.7) this takes the still simpler form

(5.19)
$$\mathcal{E}^{\infty}_{sc} = \|\hat{h}^{\infty}_{0-}\|^2_{L_2(R^3_-)}.$$

Thus the total scattered energy is just the portion of the incident signal energy associated with momenta in the half-space R^3_-; i.e., momenta directed from the sources toward the scattering layer.

Appendix 1
The Weyl-Kodaira-Titchmarsh Theory

The general Sturm-Liouville operator may be written

$$L\,\phi(y) = w^{-1}(y)\,\{-(p^{-1}(y)\phi')' + q(y)\phi\}. \tag{A.1}$$

The basic spectral theory of such operators was established by H. Weyl, K. Kodaira [10] and E. C. Titchmarsh [18]. The purpose of this appendix is to present a version of the Weyl-Kodaira-Titchmarsh theory that is applicable to the operator A_μ of this monograph.

It is true that expositions of the Weyl-Kodaira theory are available in [3,5,14,18] and a number of other textbooks and monographs. However, in these and most of the book and periodical literature, hypotheses are made concerning the form of the operator, or the continuity or differentiability of the coefficients, that limit the applicability of the theory. Thus most authors assume that $w(y) \equiv 1$ and many take $\rho(y) \equiv 1$ as well. Moreover, it is usual to assume that the coefficients are smooth functions or at least continuous. It is known that if the coefficients are sufficiently regular then L can be reduced to the Schrödinger form $L\,\phi = -\phi'' + q(y)\phi$ by changes of the independent and dependent variables [14, p. 2]. However, this technique is not applicable to operators with singular coefficients. Here a version of the Weyl-Kodaira-Titchmarsh theory is presented that is applicable to operators (A.1) with locally integrable coefficients. The concepts needed for this extension of the theory are available in the classic book of Coddington and Levinson [3].

The operator (A.1) will be studied on an arbitrary interval $I = \{y \mid -\infty \le a < y < b \le +\infty\}$. The coefficients will be assumed to have the properties

(A.2) $p(y)$, $q(y)$, $w(y)$ are defined and real valued for almost all $y \in I$,

(A.3) $p(y) > 0$ and $w(y) > 0$ for almost all $y \in I$,

(A.4) $p(y)$, $q(y)$, $w(y)$ are in $L_1^{\ell oc}(I)$,

where $L_1^{\ell oc}(I) = \{f(y) \mid f \in L_1(K) \text{ for every compact } K \subset I\}$. It is natural to study L in the Hilbert space $\mathcal{H}(I,w)$ with scalar product

$$(u,v) = \int_I \overline{u(y)}\, v(y)w(y)dy. \tag{A.5}$$

In the general theory of singular Sturm-Liouville operators two operators in $\mathcal{H}(I,w)$ are associated with L. The first is the maximal operator L_1 defined by

$$\text{(A.6)} \quad \begin{cases} D(L_1) = \mathcal{H}(I,w) \cap AC(I) \cap \{u \mid p^{-1}u' \in AC(I),\ Lu \in \mathcal{H}(I,w)\}, \\ L_1u = Lu \text{ for all } u \in D(L_1). \end{cases}$$

The second is the minimal operator L_0 defined by

$$\text{(A.7)} \quad \begin{cases} D(L_0) = D(L_1) \cap \{u \mid (L_1u,v) = (u,L_1v) \text{ for all } v \in D(L_1)\}, \\ L_0u = Lu \text{ for all } u \in D(L_0). \end{cases}$$

It can be shown that L_0 is densely defined and closed and satisfies

$$L_0 \subset L_0^* = L_1. \tag{A.8}$$

It follows that every selfadjoint realization of L in $\mathcal{H}(I,w)$ must satisfy

$$L_0 \subset L \subset L_1. \tag{A.9}$$

If $L_0 = L_0^* = L_1$ then L is said to be essentially selfadjoint. The classification of the selfadjoint realizations of L by means of boundary conditions at a and b will now be reviewed here. For essentially selfadjoint operators no boundary conditions are needed (Weyl's limit point case). The operator A_μ of §1 is essentially selfadjoint since its maximal operator is selfadjoint (cf. (3.2.1), (3.2.2)).

The Weyl-Kodaira-Titchmarsh theory provides spectral representations of the selfadjoint realizations of singular Sturm-Liouville operators. Each representation is derived from a basis of solutions of $L\psi = \lambda\psi$ and a corresponding 2×2 positive matrix measure $m(\lambda) = (m_{jk}(\lambda))$ [5, p. 1337ff]. The representation spaces are the Lebesgue spaces $L_2(\Lambda,m)$ associated with m, with norm defined by

(A.10) $$\|F\|^2_{\Lambda,m} = \int_\Lambda \sum_{j,k=1}^{2} \overline{F_j(\lambda)}\, F_k(\lambda)\, m_{jk}(d\lambda).$$

The following version of the Weyl-Kodaira-Titchmarsh theory is adapted from [5, pp. 1351-6].

<u>Theorem (Weyl-Kodaira)</u>. Let L be a selfadjoint realization of L in $\mathcal{H}(I,w)$ with spectral family $\{\Pi_L(\lambda)\}$. Let $\Lambda = (\lambda_1,\lambda_2) \subset R$ and let $\psi_j(y,\lambda)$ $(j = 1,2)$ be a pair of functions with the properties

(A.11) $\psi_j(y,\lambda) \in C(I \times \Lambda)$, $j = 1,2$,

(A.12) The pair $\psi_j(y,\lambda)$ $(j = 1,2)$ is a solution basis for $L\psi = \lambda\psi$ on I for each $\lambda \in \Lambda$.

Then there exists a unique 2×2 positive matrix measure $m = (m_{jk})$ on Λ with the following properties.

(A.13) For all $f \in \mathcal{H}(I,w)$ there exists the limit

$$\hat{f}(\lambda) = (\hat{f}_1(\lambda),\hat{f}_2(\lambda)) = L_2(\Lambda,m)\text{-}\lim_{a'\to a,b'\to b} \int_{a'}^{b'} f(y)\, (\overline{\psi_1(y,\lambda)}, \overline{\psi_2(y,\lambda)})w(y)dy.$$

(A.14) The mapping $U : \mathcal{H}(I,w) \to L_2(\Lambda,m)$ defined by $Uf = \hat{f}$ is a partial isometry with initial set $\Pi_L(\Lambda)\,\mathcal{H}(I,w)$ and final set $L_2(\Lambda,m)$.

(A.15) The inverse isomorphism of $L_2(\Lambda,m)$ onto $\Pi_L(\Lambda)\,\mathcal{H}(I,w)$ is given by

$$(U^*F)(y) = \mathcal{H}(I,w)\text{-}\lim_{\mu_1\to\lambda_1,\mu_2\to\lambda_2} \int_{\mu_1}^{\mu_2} \sum_{j,k=1}^{2} \psi_j(y,\lambda)F_k(\lambda)m_{jk}(d\lambda).$$

(A.16) For all Borel functions $\Psi(\lambda)$ on R with supp $\Psi \subset \Lambda$, one has

$$U\, D(\Psi(L)) = L_2(\Lambda,m) \cap \{\hat{f} \mid \Psi(\lambda)\hat{f}(\lambda) \in L_2(\Lambda,m)\}, \text{ and}$$

$$(U\, \Psi(L)f)(\lambda) = \Psi(\lambda)\hat{f}(\lambda).$$

Discussion of the Proof. The theorem is proved in [5] under the hypotheses $w(y) \equiv 1$, $p(y)$, $q(y) \in C^\infty(I)$ and $p(y) > 0$. To prove it under hypotheses (A.2), (A.3), (A.4) one may first prove it for the special case of the basis $\phi_j(y,\lambda)$ that satisfies $\phi_j^{(k-1)}(c,\lambda) = \delta_j^k$ where $a < c < b$. The functions $\phi_j(y,\lambda)$ are entire functions of λ and the theorem can be proved by the classical limit-point, limit-circle method of Weyl as presented in [3]. The general case can then be obtained by a change of basis from $\phi_j(y,\lambda)$ to $\psi_j(y,\lambda)$. In fact, this was the procedure used by Kodaira in his original paper [13]. The first uniqueness results for m are due to E. A. Coddington and V. A. Marčenko (see [5]). The uniqueness proof given in [5] can be extended to the case treated here.

As emphasized by Dunford and Schwartz, the utility of the Weyl-Kodaira theorem is due to the possibility of using different bases ψ_j for different portions of the spectrum of L. When a basis has been chosen one need only calculate the measure m. A general procedure for doing this, due to E. C. Titchmarsh [5, p. 1364] is known for cases in which the $\psi_j(y,\lambda)$ have analytic continuations to a neighborhood of Λ in the complex plane. However, such continuations are not always available. A procedure that is applicable when the $\psi_j(y,\lambda)$ have a one-sided continuation into the complex plane is illustrated in Chapter 3, §6 above.

Appendix 2
Stationary Phase Estimates of Oscillatory Integrals with Parameters

The method of stationary phase provides asymptotic estimates for $r \to \infty$ of oscillatory integrals of the form

$$I(r) = \int_{\mathcal{O}} \exp\{i\,r\,\theta(s)\}\, g(s)ds, \quad \mathcal{O} \subset R^m.$$

The method was needed in Chapter 4, §4 to estimate the oscillatory integral in equation (4.4.3). There the integrand g and phase θ contain parameters and estimates are needed that are uniform in these parameters, at least on compact sets. This suggests the study of oscillatory integrals of the form

$$I(r,\xi) = \int_{\mathcal{O}} \exp\{i\,r\,\theta(s,\xi)\}\, g(r,s,\xi)ds \tag{1}$$

where $r > 0$, $s \in R^m$ and $\xi \in R^n$. $\theta(s,\xi)$ is a real-valued phase function and it is assumed that there are open sets $\mathcal{O} \subset R^m$ and $\mathcal{O}' \subset R^n$ such that

$$D_s^\delta\, \theta(s,\xi) \in C(\mathcal{O} \times \mathcal{O}') \text{ for all multi-indices } \delta, \tag{2}$$

$$D_s^\delta\, g(r,s,\xi) \in C(R_+ \times \mathcal{O} \times \mathcal{O}') \text{ for all multi-indices } \delta, \tag{3}$$

where R_+ denotes the positive real numbers. Moreover, it is assumed that there is a compact set $K \subset \mathcal{O}$ such that

$$\text{supp } g \subset R_+ \times K \times \mathcal{O}'. \tag{4}$$

Estimates of $I(r,\xi)$ for $r \to \infty$ are sought which are uniform in ξ on compact subsets of $\mathcal{O}'$. For large values of r the exponential in (1) is highly oscillatory except near critical points of the phase function $\theta(s,\xi)$. Two estimates of $I(r,\xi)$ are given here, corresponding to the cases of no critical points and one critical point, respectively. The first case is formulated as

Theorem A.1. Assume that

$$\nabla_s \theta(s,\xi) = (\partial\theta/\partial s_1,\cdots,\partial\theta/\partial s_m) \neq 0 \text{ for all } (s,\xi) \in K \times \mathcal{O}'. \tag{5}$$

Moreover, assume that for each compact set $K' \subset \mathcal{O}'$, each $r_0 > 0$ and each positive integer k there exists a constant $M = M(K,K',r_0,k) > 0$ such that

$$|D_s^\delta g(r,s,\xi)| \leq M \text{ for all } r \geq r_0,\ s \in K,\ \xi \in K' \text{ and } |\delta| \leq k. \tag{6}$$

Then there exists a constant $C = C(K,K',r_0,k,g) > 0$ such that

$$|I(r,\xi)| \leq C\, r^{-k} \text{ for all } r \geq r_0 \text{ and } \xi \in K'. \tag{7}$$

In the second case considered here $\theta(s,\xi)$ has a unique non-degenerate critical point $s = \tau(\xi)$ for each $\xi \in \mathcal{O}'$. It is formulated as

Theorem A.2. Assume that there is a function $\tau \in C^\infty(\mathcal{O}',\mathcal{O})$ such that, for all $\xi \in \mathcal{O}'$,

$$\nabla_s \theta(s,\xi) = 0 \text{ if and only if } s = \tau(\xi). \tag{8}$$

Moreover, assume that the Hessian $\theta''(s,\xi) = (\partial^2\theta(s,\xi)/\partial s_j \partial s_k)$ satisfies

$$\det \theta''(\tau(\xi),\xi) \neq 0 \text{ for all } \xi \in \mathcal{O}'. \tag{9}$$

In addition, assume that for each compact set $K' \subset \mathcal{O}'$ and each $r_0 > 0$ there exists a constant $M = M(K,K',r_0)$ such that

$$|D_s^\delta g(r,s,\xi)| \leq M \text{ for all } r \geq r_0,\ s \in K,\ \xi \in K' \text{ and } |\delta| \leq m + 5. \tag{10}$$

Then there exists a constant $C = C(K,K',r_0,g) > 0$ such that if $q(r,\xi)$ is defined by

$$I(r,\xi) = (2\pi)^{m/2} \frac{\exp\{i\, r\, \theta(\tau(\xi),\xi) + \frac{\pi}{4} \operatorname{sgn} \theta''(\tau(\xi),\xi)\} g(r,\tau(\xi),\xi)}{r^{m/2}\ |\det \theta''(\tau(\xi),\xi)|^{1/2}} + q(r,\xi) \tag{11}$$

then

(12) $$|q(r,\xi)| \leq C\, r^{-m/2-1} \quad \text{for all } r \geq r_0 \text{ and } \xi \in K'.$$

In (11), sgn $\theta''(s,\xi)$ is the signature of the real symmetric matrix $\theta''(s,\xi)$.

The uniform estimates given above are due, in essence, to M. Matsumura [15]. No proofs are offered because the theorems can be proved by following Matsumura's proofs and recognizing that under the hypotheses formulated above his estimates are uniform for $\xi \in K'$.

References

[1] Agmon, S. Lectures on Elliptic Boundary Value Problems. Van Nostrand, 1965.

[2] Brekhovskikh, L. M. Waves in Layered Media. Academic Press, 1960.

[3] Coddington, E. A. and Levinson, N. Theory of Differential Equations. McGraw-Hill, 1955.

[4] Dermenjian, Y., Guillot, J. C. and Wilcox, C. H. Comportement asymptotique des solutions de l'équation des ondes associée à l'opérateur d'Epstein. C. R. Acad. Sci. Paris 287, 731-734 (1978).

[5] Dunford, N. and Schwartz, J. T. Linear Operators Part II: Spectral Theory. Interscience, 1963.

[6] Friedlander, F. G. Sound Pulses. Cambridge Univ. Press, 1958.

[7] Guillot, J. C. and Wilcox, C. H. Spectral analysis of the Epstein operator. Univ. of Utah Technical Summary Rept. #27, 1975.

[8] Guillot, J. C. and Wilcox, C. H. Spectral analysis of the Epstein operator. Proc. Roy. Soc. Edinburgh 80A, 85-98 (1978).

[9] Hartman, P. Ordinary Differential Equations. Wiley, 1964.

[10] Hörmander, L. Linear Partial Differential Operators. Springer-Verlag, 1963.

[11] Kato, T. Perturbation Theory of Linear Operators. Springer-Verlag, 1966.

[12] Kneser, A. Untersuchungen über die reelen Nullstellen der Integrale linearer Differentialgleichungen. Math. Ann. 42, 409-435 (1893).

[13] Kodaira, K. The eigenvalue problem for ordinary differential equations of the second order and Heisenberg's theory of S-matrices. Amer. J. Math. 71, 921-945 (1949).

[14] Levitan, B. M. and Sargsjan, I. S. Introduction to Spectral Theory, AMS Translations of Mathematical Monographs V. 39, American Mathematical Society, Providence 1975.

[15] Matsumura, M. Asymptotic behavior at infinity for Green's functions of first order systems with characteristics of nonuniform multiplicity. Publ. RIMS, Kyoto Univ. 12, 317-377 (1976).

[16] Naimark, M. A. Linear Differential Equations Part II. Ungar, 1967.

[17] Pekeris, C. L. Theory of propagation of explosive sound in shallow water. Geol. Soc. Am., Memoir 27 (1948).

[18] Titchmarsh, E. C. Eigenfunction Expansions, V. I. OUP, 1946.

[19] Tolstoy, I. The theory of waves in stratified fluids including the effects of gravity and rotation. Rev. Mod. Phys. 35, 207-230 (1963).

[20] Tolstoy, I. and Clay, C. S. Ocean Acoustics. McGraw-Hill, 1966.

[21] Weyl, H. Über gewöhnliche Differentialgleichungen mit Singularitäten und die zugehörigen Entwicklungen willkürlicher Funktionen. Math. Ann. 68, 220-269 (1910).

[22] Wilcox, C. H. Initial boundary value problems for linear hyperbolic partial differential equations of the second order. Arch. Rational Mech. Anal. 10, 361-400 (1962).

[23] Wilcox, C. H. Transient electromagnetic wave propagation along a dielectric-clad conducting plane, Univ. of Utah Technical Summary Rept. #22, May 1973.

[24] Wilcox, C. H. Transient acoustic wave propagation in a symmetric Epstein duct, Univ. of Utah Technical Summary Rept. #25, May 1974.

[25] Wilcox, C. H. Scattering Theory for the d'Alembert Equation in Exterior Domains. Springer Lecture Notes in Mathematics, V. 442. Springer-Verlag, 1975.

[26] Wilcox, C. H. Spectral analysis of the Pekeris operator. Arch. Rational Mech. Anal. 60, 259-300 (1976).

[27] Wilcox, C. H. Transient electromagnetic wave propagation in a dielectric waveguide. Istituto Nazionale di Alta Matematica, Symposia Mathematica XVIII, 239-277 (1976).

[28] Wilcox, C. H. Spectral and asymptotic analysis of acoustic wave propagation. Boundary Value Problems for Linear Evolution Partial Differential Equations. H. G. Garnir, Ed. Reidel, 1977.

[29] Wilcox, C. H. Transient acoustic wave propagation in an Epstein duct. Univ. of Utah Technical Summary Report #36, November 1979.

Index

Applied Mathematical Sciences

cont. from page ii

36. Bengtsson/Ghil/Källén: **Dynamic Meterology: Data Assimilation Methods.**
37. Saperstone: **Semidynamical Systems in Infinite Dimensional Spaces.**
38. Lichtenberg/Lieberman: **Regular and Stochastic Motion.** (cloth)
39. Piccinini/Stampacchia/Vidossich: **Ordinary Differential Equations in R^n.**
40. Naylor/Sell: **Linear Operator Theory in Engineering and Science.** (cloth)
41. Sparrow: **The Lorenz Equations: Bifurcations, Chaos, and Strange Attractors.**
42. Guckenheimer/Holmes: **Nonlinear Oscillations, Dynamical Systems and Bifurcations of Vector Fields.**
43. Ockendon/Tayler: **Inviscid Fluid Flows.**
44. Pazy: **Semigroups of Linear Operators and Applications to Partial Differential Equations.**
45. Glashoff/Gustafson: **Linear Optimization and Approximation: An Introduction to the Theoretical Analysis and Numerical Treatment of Semi-Infinite Programs.**
46. Wilcox: **Scattering Theory for Diffraction Gratings.**
47. Hale et al.: **An Introduction to Infinite Dimensional Dynamical Systems — Geometric Theory.**
48. Murray: **Asymptotic Analysis.**
49. Ladyzhenskaya: **The Boundary-Value Problems of Mathematical Physics.**
50. Wilcox: **Sound Propagation in Stratified Fluids.**
51.. Golubitsky/Schaeffer: **Bifurcation and Groups in Vifurcation Theory, Vol. I.**
52. Chipot: **Variational Inequalities and Flow in Porous Media.**
53. Majda: **Compressible Fluid Flow and Systems of Conservation Laws in Several Space Variables.**